Historical Scientific Instruments in Contemporary Education

Scientific Instruments
and Collections

STUDIES PUBLISHED UNDER THE AUSPICES OF THE
SCIENTIFIC INSTRUMENT COMMISSION

General Editor

Giorgio Strano, *Museo Galileo, Istituto e Museo di Storia della Scienza, Florence*

Editorial Board

Mara Miniati, *Museo Galileo, Istituto e Museo di Storia della Scienza, Florence*
Alison Morrison-Low, *National Museums Scotland, Edinburgh*
Sara J. Schechner, *Harvard University, Cambridge, MA*

VOLUME 9

The titles published in this series are listed at *brill.com/sico*

Contents

Foreword   VII
List of Illustrations   X
Notes on Contributors   XVI

Introduction: Using Historical Scientific Instruments in Contemporary Education
Experiences and Perspectives   1
   Elizabeth Cavicchi and Peter Heering

1   Reading Instruments for Historical Scientific Practice
    An Experiential Pedagogy for Material Culture   14
       Alistair Kwan

2   Filming Nineteenth Century Physics Demonstrations with Historical Instruments   34
       Paolo Brenni

3   Making It about the Objects: A Reboot of a History of Science Course   50
       Janet Laidla

4   Using Original Instruments from a Museum Collection in Demonstrations   65
       Jan Waling Huisman

5   The Collections of Scientific Instruments of the Faculty of Sciences of Rennes
    A Tool for School Education and for the Training of Students and Teachers   84
       Julie Priser and Dominique Bernard

6   The Collection of Scientific Instruments from the Maraslean Teaching Center and Experimental Science Education: Then and Now   105
       Panagiotis Lazos, Constantina Stefanidou and Constantine Skordoulis

CONTENTS

7   Examples of the Use in Education of Historical Physics Instruments at Secondary School and University Level in France Supported by ASEISTE   122
    Françoise Khantine-Langlois, Alfonso San-Miguel and Pierre Lauginie

8   The Use of the Museum Collection for Educational Purposes   139
    Roland Carchon and Danny Segers

9   Historical Scientific Instruments in Exploratory Teaching and Learning   158
    Elizabeth Cavicchi

10   "What Is Happening in the Lab?" Transforming the School Laboratory into a Contextual Science Teaching Environment   181
    Flora Paparou

11   Historical Instruments, Education, and Do-It-Yourself in the Cabinet of Curiosity of Brest, France
    University Experiences in Mathematics   209
    Frédérique Plantevin and Pietro Milici

12   Educational Experiences in Re-enacting Historical Experimental Procedures   226
    Peter Heering

13   The Lorentz Lab: Reviving the Scientific History of Teylers Museum with Working Replicas   243
    Trienke M. van der Spek

14   The Fall of Bodies according to Galileo
    A Free Adaptation from the Geneva Museum of the History of Science   263
    Stéphane Fischer

    Index   281

Printed in the United States
By Bookmasters

Foreword

In the last twelve years, nine volumes of the *Scientific Instruments and Collections* series have appeared. They include contributions by over ninety different authors working as university scholars, museum directors, collection managers, curators, restorers, collectors, antiquarians and, with this volume, science and mathematics educators.

Over such a time interval, several changes have occurred. Sadly, a few authors have passed away, and I wish to recall Inge Keil, Inga Elmqvist Söderlund and James Caplan. On the positive side, *Scientific Instruments and Collections* has been transformed from a branch of the *History of Science and Medicine Library* into an independent series. I therefore would like to thank Stefan Einarson and Rosanna Woensdregt of Brill for the trust accorded to the Scientific Instrument Commission (SIC) of the International Union of History and Philosophy of Science and Technology (IUHPST), the volumes editors and authors, the series editorial board and myself. We always try to publish only the most recent research on ancient and historical scientific instruments.

The present volume is, in fact, the first book completely dedicated to the use of historical scientific instruments in educational activities for primary and secondary schools, colleges and universities, museums, exhibitions and cultural festivals. As Elizabeth Cavicchi and Peter Heering specify in the *Introduction*, there is no intention to systematise every aspect of the topic. The volume presents selected papers from annual SIC meetings, and the authors represent nine countries (Belgium, Estonia, France, Germany, Greece, Italy, Switzerland, The Netherlands and the USA) and about twenty institutions. Nevertheless, the educational initiatives expounded upon in the following pages outline a vast array of combinations between historical scientific instruments, educators, students and the general public. The connecting live-wire is the conviction that historical apparatus facilitates the understanding of those scientific principles which usually disappear within the "black boxes" of today's laboratories.

Among the collection of examples to adopt and adapt to any exigencies, the readers might, however, find too much. In some cases, an educational activity presented as regular practice by an author cannot be identically replicated elsewhere. The public use of historical scientific instruments is limited by the law of some countries and the internal protocols of different institutions. Such limits are only partially emphasised in the volume chapters, and require a few words of caution.

Firstly, the law and safety protocols might regulate – limit or forbid – the public use of some objects (especially those with sharp parts), dangerous substances (for example: mercury, asbestos and other chemicals), and physical phenomena (electricity, radioactivity, and so on). Even the best precautions presented by an author might not be enough when moved to another context. Before replicating any activities, it is therefore important to carefully check if they are permitted by local law.

Secondly, the concept of "cultural heritage" is still evolving and strongly country-dependent. The local law and regulations might establish if a scientific instrument is a "cultural heritage item", and if and how it can be touched, activated, used or restored. For example, we learn from the book that MIT students visiting Harvard University were allowed to handle an original Galileo compass. At the Museo Galileo in Florence, access to another original Galileo compass would be granted only to students with specific university-level research projects, and under very strict controls. As another example, we learn that, in France, science students restore historical scientific instruments. But restoration is "cultural heritage"-dependent too. While an "old" scientific instrument in private hands can be restored by anyone, in Italy an "historical" scientific instrument in a public institution can only be restored by a professional restorer and his/her pupils. As a final example, we learn from the book about historical instruments put into operation. Once again, this depends on the law, as repeated public performances favours the deterioration of the item. In this case, the authors of the book offer different perspectives, from the regular use of not-so-important devices, through the exceptional use of important devices, up to the use of museum devices activated by the expert in order to finally film their functioning once and for all.

The invitation to attentively examine local law before trying to emulate anything, does not emphasise a limit, but rather an interesting aspect of the book. The volume takes a picture of the awareness of the relationship between historical scientific instruments and the "cultural heritage" in different countries and institutions. The purpose of the many educational activities presented is, by the way, to increase such an awareness by the interaction between the students and scientific instruments from the past.

To conclude, at the end of twelve years between the first and the ninth volume of this series, another important change will take place. From the next volume, the tenth, *Scientific Instruments and Collections* will pass to the hands of another very capable general editor, who will bring in new ideas and indicate new perspectives. It is my pleasure to wish A.D. Morrison-Low success in the future; and, in addition to hers, acknowledge the steady collaboration that

I received from the present and past series editors: Stephen Johnston, Mara Miniati and Sara Schechner. Many thanks to them and to all the volume editors and authors who contributed to the success of the series.

Giorgio Strano (General Editor)
May 2021

Illustrations

Figures

1.1   Philippe Starck, Juicy Salif, 1990 (Photo by Jonas Forth, 2013)   18

1.2   Some of Kamprani's uncomfortable objects (Photo by Simon Berry, *Bad Design*)   20

1.3   Eighteenth-century style octant (W.H. Hall (ed.), *The New Royal Encyclopaedia*, London, 1791)   22

1.4   Sextant and its use in a twentieth-century diagram (C. Fisher, J.H. Gerould and J.P. Poole, *The Marvels and Mysteries of Science*, New York, 1941)   22

1.5   Sextant and its use in a late nineteenth-century depiction (A. Guillemin, *El mundo físico*, Barcelona, 1883)   23

1.6   Equatorial telescope at Greenwich (*Leisure Hour* 525, 1862)   25

1.7   Zenith telescope in the field (Canada Dept. of Mines and Technical Surveys / Library and Archives Canada)   26

1.8   Spinthariscope illustrations (*Scientific American* 91 (1904); *Dundee Evening Post*, 1903, © The British Library Board, used with permission via the British Newspaper Archive; H. Geiger and E. Marsden, "On a Diffuse Reflection of the α-Particles", *Proceedings of the Royal Society A* 82, 1909)   27

2.1   A. Giatti is ready to show Chladni's figures while A. Chiavacci is filming   35

2.2   An engraving illustrating how to blow up a metallic wire with electricity (A. Guillemin, *Les phénomènes de la physique*, Paris, 1868)   40

2.3   P. Brenni is about to explode a metallic wire   41

2.4   P. Brenni and A. Giatti prepare Duboscq's lantern with the arc light and the dissolving view apparatus   43

2.5   P. Brenni and R. Wittje experimenting with an electric arc   44

2.6   A nineteenth-century engraving illustrating how to use a spectroscope (A. Guillemin, *Les phénomènes de la physique*, Paris, 1868)   45

2.7   P. Brenni using a nineteenth-century spectroscope   46

2.8   A frame of the video dedicated to the Faraday effect   49

3.1   Craniometer or, according to the Baltic-German anatomist E. Landau, a cephalograph (University of Tartu Museum; Photo by Mart Küng)   55

3.2   L. Ombrédanne's anaesthesia mask by Mathieu & Gentile Collin and H. Windler (University of Tartu Museum; Photo by Gert Klaasen)   56

4.1   Air pump by J. van Musschenbroek, Leyden, 1698 (University Museum Groningen; Photo by S.L. Ackermann)   66

4.2   Excerpt from *Series Lectionum 1697* (Special Collections University Library Groningen)   67

ILLUSTRATIONS

4.3 The brachistochrone (W.J.'s Gravesande, *Physices elementa mathematica...*, Leiden, 1742) 71

4.4 Universal kymograph (E. Zimmermann, *Wissenschaftlische Apparate: Liste 50*, Leipzig, 1928; University Museum Groningen) 74

4.5 Electric motor by Stratingh and Becker, 1835, and its replica (University Museum Groningen and private collection; Photo by S.L. Ackermann) 76

4.6 Ranschburg apparatus for testing perception (E. Zimmermann, *Wissenschaftlische Apparate: Liste 50*, Leipzig, 1928; University Museum Groningen) 78

4.7 Ribokov Cage, unknown maker, third quarter of twentieth century (Collections Archief en Documentatiecentrum Nederlandse Gedragswetenschappen – ADNG; Photo by S.L. Ackermann) 80

4.8 Vacuum pump, E. Leybold's Nachfolger Cöln-Rhein, 1905 (University Museum Groningen; Photo by S.L. Ackermann) 81

4.9 Experiment for testing memory and perception (Photo by the Photographic Services Department of the University of Groningen) 82

5.1 Internship students N. Rozé and J. Thouin operate a double Helmholtz siren; University of Rennes 1, 2005 86

5.2 The QUESACO project with Koenig's sound analyser; University of Rennes 1, 2017 92

5.3 W. Tobin and A. Faisant reconstruct the experiment of the Foucault pendulum, Rennes 93

5.4 Pupils and teachers in the instruments gallery; University of Rennes 1, 2014 96

5.5 When pupils are experimenting! Rennes, 2019 96

5.6 One poster of public demonstrations: Journée Européenne des Collections Universitaires; University of Rennes 1, 2014 98

5.7 The experiment of P. and M. Curie reconstructed with quartz; University of Rennes 1, 2015 100

5.8 Exhibition of instruments from chemical laboratories at the Ecole Nationale Supérieure de Chimie de Rennes, 2019 101

6.1 The Maraslean Teaching Center around 1930 (Photo by Petros Poulidis ERT-Archives) 107

6.2 Experiments with hydrostatic balances in the Maraslean Teaching Center (Yiorgos Palaiologos, *The institution of pedagogical academies ...*, 1939) 107

6.3 A team of students study a Müller apparatus (Photo by P. Lazos) 110

6.4 The Wimshurst machine after the repair (Photo by P. Lazos) 112

6.5 The three pendulums after the repair (Photo by P. Lazos) 113

6.6 The instruments in the laboratory (Photo by P. Lazos) 114

6.7 A team of students trying to conceptualise how the acoustic siren works (Photo by P. Lazos) 116

6.8	Students studying the mercury fine shower apparatus (Photo by P. Lazos)	117

7.1 Example from the *Encyclopédie des instruments de l'enseignement de la Physique...*: instrument to demonstrate the properties of the cycloid 124

7.2 Demonstration of physics at the Lycée Hoche of Versailles, postcard, beginning of the twentieth century (*Encyclopédie des instruments de l'enseignement de la Physique...*, ASEISTE, 2016) 125

7.3 Baroscope from the secondary school Collège du Château at Morlaix (Photo J.-Y. Blaise) 127

7.4 Physics cabinet at the Lycée Hoche of Versailles, postcard, beginning of the twentieth century (*Encyclopédie des instruments de l'enseignement de la Physique...*, ASEISTE, 2016) 129

7.5 The Archimedes screw from the Lycée Hoche of Versailles (Photo J. Millet) 131

7.6 Poster of the exhibition *Histoires de l'eau*, Musée du Lycée Hoche, Versailles, September 2019–December 2020 132

7.7 Master students from University Claude Bernard Lyon 1 restoring a Zénobe Gramme machine 134

7.8 Master students from University Claude Bernard Lyon 1 (UCBL) presenting restored physics instruments 135

8.1 Anorthoscope disk of Plateau (Courtesy Ghent University Museum) 144

8.2 Slide explaining the functioning of the anorthoscope of Plateau (Courtesy Ghent University Museum) 145

8.3 One-page leaflet on the bioscope, printed by Duboscq, Paris, 1853 (Courtesy Ghent University Museum) 147

8.4 Disc of Duboscq (Courtesy Ghent University Museum) 148

8.5 Replica of bioscope of Duboscq (Courtesy Ghent University Museum) 149

8.6 Bench of Melloni for infrared research (Courtesy Ghent University Museum) 151

8.7 One of the three objects prepared by L. Baekeland for F. Swarts (Courtesy Ghent University Museum) 153

8.8 The original van Leeuwenhoek microscope from the collection "Stad Antwerpen" (Courtesy Ghent University Museum) 155

9.1 Students explore an astrolabe replica; disassemble a vacuum tube housing; examine the Sperry revolving mirror and catch its reflections on paper 162

9.2 Microscope gallery visit at the MIT Museum; and blow-piping with Bunsen burner 164

9.3 Three microscope drawings by G. Antunes, S. Kiley and L. Rodriguez, 2007 165

9.4 Students visit the MIT Museum's off-site storage 168

9.5 Students visit the MIT Museum's off-site storage 170

9.6 Sighting and calculating with a laser-cut Galileo compass 172

ILLUSTRATIONS

9.7 Galileo's geometrical and military compass, Padua, c. 1604 (Harvard Collection of Historical Scientific Instruments, Cambridge, MA, inv. no. DW0950) compared with laser-cut model 173

9.8 F. Liuni's section architectural drawing of her astrolabe museum (F. Liuni, *Experiencing Mathematical Proves* ..., Master Thesis, Cambridge, MA, 2016) 176

9.9 F. Liuni explains her exhibit, *Syntax of an Astrolabe*, Foyer Gallery, Harvard, Cambridge, MA, 2017 177

9.10 Laser-cut astrolabe in class; Gary Stilwell's stereographic projections corresponding to astrolabe plates for Mercury and Venus 178

10.1 Athens Science Festival 2016. Re-enactment of a Royal Society meeting and the Magdeburg experiment 183

10.2 Experiments on falling objects and scheme of a switch to check the fall in the air 192

10.3 Tools for replicating Berti's experiment and original sketches of the experiment (E. Maignan, *Cursus Philosophicus*, Toulouse, 1653; R. Magiotti, "Letter to Mersenne", 1648) 193

10.4 Large surface conductors and Leyden jars to strengthen the Wimshurst machine spark 196

10.5 Electromagnetism. Writing with electricity; Adding a magnetic needle to illustrate the voltage change; and Lissajous curves with a Braun tube 197

10.6 Science theatre, 2015, and lecture demonstrations at the Athens University History Museum, 2014–2015 200

10.7 The escape-room project. Students interpret J. Tyndall and Lord Rayleigh; Visitors explore the navigation process at the Athens Science Festival 2016 202

10.8 A view of the Flying Machine section at the Deutsches Museum, Munich (2017) 204

10.9 Front cover of a notebook from Dachau (courtesy of Yad Vashem Archives, Jerusalem) 205

11.1 The Cabinet of Curiosity of Brest (Photo by M. Danaux, UBO) 210

11.2 Mathematical instruments of the Institute for Research in the Teaching of Mathematics 210

11.3 Posters of the exhibitions loaned by the Musée des arts et métiers and the Science in the Seine and Heritage Association, or produced by IREM 211

11.4 Examples of works in the Cabinet of Curiosity 212

11.5 Different steps of the building processes in the Cabinet of Curiosity 216

11.6 The "parabolograph" and Perks' machine 220

11.7 Realisation of a "parabolograph" 221

11.8	The exponential machine 222
12.1	Three electrical 'toys': electrical boxer, electric hailstorm and electrical huntsman (© Europa-Universität Flensburg) 228
12.2	Two electrostatic generators (© Europa-Universität Flensburg) 229
12.3	Plate with electrically charged person attracting light objects (J.T. Desaguliers, *De Natuurkunde Uit Ondervindingen*, Amsterdam, 1751; Reproduced with permission of the Landesbibliothek Oldenburg) 231
12.4	Electrically charged person attracting light objects (Photo by C. Anrich, Phänomenta Flensburg) 232
12.5	Boxplot for the evaluation of the electrical Salon 235
12.6	Boxplot for analysis of visitors' satisfaction 235
12.7	Two *cameras obscura* used in Projekt Galilei (© Europa-Universität Flensburg) 238
12.8	Teacher working on an astatic galvanometer (Photo by M. Engel, Europa-Universität Flensburg) 239
13.1	The large electrostatic generator (M. van Marum, *Beschryving eener ongemeen groote Elektrizeer-Machine...*, Haarlem, 1785) 245
13.2	The 1885 instrument room with van Marum's large electrostatic generator (Photo Teylers Museum) 246
13.3	The Lorentz Lab (Photo Teylers Museum) 249
13.4	The working replica of van Marum's machine (Photo Teylers Museum) 251
13.5	Working replicas of Spiral by Riess; Ritchie motor; Faraday's ring; and Clarke dynamo (Photos A. Stoelwinder) 254
13.6	Working replicas of Sine galvanometer; Ampère's Law Instrument; Oersted demonstration instrument; and Voltaic pile (Photos Anton Stoelwinder) 255
13.7	Barlow's wheel replica, using Galinstan in the trough (Photo Teylers Museum) 256
13.8	School group experimenting at the Lorentz Lab (Photo Teylers Museum) 258
14.1	Replica of the hemispheres of Magdeburg presented outside of the Museum, Geneva, 2010 (© Musée d'histoire des sciences de Genève) 264
14.2	First page of a re-edition of Galileo's *Discourses* published in Bologna in 1655 (© Musée d'histoire des sciences de Genève) 265
14.3	The inclined plane of the Museum (© Musée d'histoire des sciences de Genève) 270
14.4	Detail of a bell installed on the inclined plane (© Musée d'histoire des sciences de Genève) 271
14.5	Demonstration of the accelerated movement of the balls rolling along the inclined plane (© Musée d'histoire des sciences de Genève) 272
14.6	The volunteer experimenter concentrates on releasing the ball at the beat of the electronic metronome (© Musée d'histoire des sciences de Genève) 274

ILLUSTRATIONS

14.7 Demonstration of the law of the pendulum (© Musée d'histoire des sciences de Genève)   276

14.8 Demonstration of falling bodies in vacuum with the Newton tube (© Musée d'histoire des sciences de Genève)   277

Table

8.1 Bachelor projects realised in the University of Ghent (UGhent) Museum   142

Notes on Contributors

Dominique Bernard

was a "Maître de Conférences" in physics at the University of Rennes 1, France, where he rescued old scientific instruments. He works on the history of ancient physics instruments, developing an experimental approach to them. In 2018, he published a book on the University collection, which won the 2019 Scientific Information Prize of the Academy of Sciences.

Paolo Brenni

studied experimental physics at the University of Zürich, where he graduated in 1981. He then specialised in the history of scientific instruments and of precision industry in the period from the beginning of the eighteenth century to the mid-twentieth century. He is a researcher in Florence for the Italian National Research Council (CNR) and collaborates with the Museo Galileo. He catalogued and restored several collections of scientific instruments in Italy and abroad, and has written numerous articles about instrument history, their trade and their production. Since 2005, he has been President of the Scientific Instrument Society.

Roland Carchon

holds a PhD from Ghent University in nuclear physics. During his professional life he was a researcher at the Nuclear Research Centre (Mol-Belgium) and the International Atomic Energy Agency (IAEA-Vienna). After retiring, he has been a collaborator at Ghent University Museum for the History of Sciences, with interests in the educational applications of museum collections in a historical context and the popularization of scientific theories.

Elizabeth Cavicchi

completed a doctor of education (EdD) degree at Harvard University; master's degrees at Harvard, Boston University and MIT; undergraduate degrees at MIT. She has written and presented internationally on explorations interweaving history, science phenomena, teaching and learning. At MIT's Edgerton Center, Cavicchi encourages learners to be explorers. Her seminars provide direct experiences with observation, experiment, instruments, history and social justice. Cavicchi's artwork spans watercolours, pastels and sculptural media.

Stéphane Fischer

is assistant curator at the Museum of the History of Science in Geneva, Switzerland, and is in charge of the Museum's collections. He organises and sets up numerous projects – exhibitions, replicas, demonstrations, publications – in connection with the collections and their promotion to the public.

Peter Heering

is professor of physics and its didactics at the Europa-Universität Flensburg, Germany. His research focuses on the history of physics, especially experimental practice, which he investigates using the replication method, the use of historical content in science education, and the historical development of teaching experiments in physics education.

Jan Waling Huisman

holds a BASc in Environmental Sciences and Engineering, and studied as a physics teacher. A staff member of the University Museum Groningen, in the Netherlands, since 1989, he is a collections manager with curatorial tasks. He has cooperated in dozens of exhibitions and projects in different roles, including project management and designing, and engineering interactives. Using old instruments to engage the public in understanding science is one of his key targets.

Françoise Khantine-Langlois

PhD and former professor in the technical department of the University Claude Bernard Lyon 1, is currently an associate researcher at the Sciences and Society, Historicity, Education, Practices (S2HEP) laboratory at Lyon 1 University. She manages the University's collection of physical instruments and is president of the Association de Sauvegarde et d'Étude des Instruments Scientifiques et Techniques de l'Enseignement (ASEISTE).

Alistair Kwan

studied and taught physics, history of science, and education in the United States, Australia and New Zealand universities. He has worked on how historical objects, environments and architecture can constitute primary source evidence in education, research, and heritage interpretation, especially for voices and kinds of knowledge that the textual record does not represent.

Janet Laidla

has a PhD in history. Her research has previously concentrated on early modern historiography in Estonia and Livonia. Her current research interests lie in the history of knowledge of the modern period. She currently works at the University of Tartu as Lecturer of Estonian History and Curator at the University of Tartu Museum.

Pierre Lauginie

is a former lecturer and researcher in physics, who has developed an experimental approach to history of science based on adaptations of historical experiments. His present interests concern the history of instruments and measurement, and the popularization of science.

Panagiotis Lazos

is a physicist with an MSc in History and Philosophy of Science from the Technical University of Athens and the National and Kapodistrian University of Athens. He is a PhD candidate in the latter. He had taught high school physics for more than fifteen years and is currently the Head of the 4th Laboratory Center of Natural Sciences of Athens. His research and publications are on the history of scientific instruments, history of science, didactics of science, and the use of open source platforms in science education.

Pietro Milici

has a PhD in mathematics (University of Palermo, Italy) and in epistemology (Paris-Sorbonne University, France). He is a researcher in the Department of Theoretical and Applied Sciences (DiSTA) of the University of Insubria (Varese, Italy). He is the founder of www.machines4math.com, a Research and Development company for tangible educational materials. He collaborates with the Cabinet of Curiosity of the University of Brest, France; is a component of the EuroPoleni research group (on the eighteenth-century Italian polymath Giovanni Poleni). His main research interests deal with mathematical machines (mainly for tractional motion) from historical, philosophical, and educational perspectives.

Flora Paparou

works currently as a science teacher in secondary education, in Athens, Greece. She is a chemical engineer and holds a PhD in science education. Her research focuses on the material culture of science, as well as on the integration of the history of science in science teaching. From 2003 to 2008 she organised the Science Museum of Chios educational programme. Since 2012, she has been systematically involved in the documentation of the Athens University scientific instrument collections.

Frédérique Plantevin

is lecturer in mathematics at the University of Brest, France, and member of the Laboratorie de Mathematiques de Bretagne Atlantique (LMBA). She is involved in initial and continuing teachers' training, in particular through her implication in the Instituts de Recherche sur l'Enseignement des Mathématiques (IREM) network. She has developed a line of work with primary and secondary teachers on historical instruments in the classes. In 2016, she founded the Cabinet of Curiosity in the Faculty of Science where the collection of outdated scientific instruments is housed.

Julie Priser

is collection assistant and PATSTEC (Patrimonie Scientifique et Technique Contemporaine) project manager for the Brittany region. She works for the protection, conservation, and valorization of old and contemporary scientific instruments at the University of Rennes 1, France.

Alfonso San-Miguel

is physics professor at the University Lyon 1, Director of the Ampère Physics Federation, and president of the Rhône region branch of the French Physical Society. He is also member of the Friends of Ampère Society. He has initiated outreach projects involving university students and the above mentioned learned societies. He has also set up projects for the safeguarding and development of the scientific heritage with physics master's students in collaboration with the Association de Sauvegarde et d'Étude des Instruments Scientifiques et Techniques de l'Enseignement (ASEISTE).

Danny Segers

holds a PhD from Ghent University in materials research by nuclear methods. He has been a professor of physics with teaching duties in the departments of informatics and veterinary medicine. In the period from 2006 till 2016 he was the director of the Ghent University Museum for the History of Science. During that time, he was teaching a course on the history of science.

Constantine Skordoulis

is Professor of Epistemology and Didactical Methodology of Physics at the Department of Primary Education, National and Kapodistrian University of Athens. He is the academic coordinator of the postgraduate program "Secondary Science Teachers Education" at the Hellenic Open University. He studied natural sciences at the University of Kent at Canterbury, UK, and worked as a visiting researcher at the Universities of Oxford and Groningen. His research interests include the history of science and science education from a critical perspective.

Trienke M. van der Spek

is chief curator and head of the science collections at Teylers Museum, Haarlem. She previously worked at Rijksmuseum Boerhaave in Leiden as curator and head of collections. Graduating as a chemist, she also held positions in science education at Nemo Science Museum in Amsterdam and at the University of Amsterdam. Her research interests include the popularisation of science in the nineteenth century and Teylers Museum's institutional and collection history.

Constantina Stefanidou

is a physicist with a PhD in Science Education from University of Athens. After a long period of teaching science in secondary education, she now has a position in the Department of Education at the National and Kapodistrian University of Athens. Her research and publications are on didactics of science, focusing on historical and philosophical perspectives as well as conceptual difficulties and their relation to model-based teaching and learning, and informal science education. She participates in international conferences (ESERA, IHPST, etc.) and science communication actions.

INTRODUCTION

Using Historical Scientific Instruments in Contemporary Education

Experiences and Perspectives

Elizabeth Cavicchi and Peter Heering

From historical to contemporary times, education and science have undergone shifts, reversals, changes and transformations along multiple dimensions including: makeup of participants, means of practice, and matters and perspectives for study and research. While local circumstances contribute, some trends are evident. Participants in many settings of education and science today include those not openly welcome under past restrictions regarding age, class, disability, ethnicity, gender, gender nonconforming identity, economic status, race and other categories. Increasingly, means of doing education and science involve globally connecting networks where communication and information retrieval is often immediate. Matters and perspectives in present-day education and science involve interaction and participation in culturally-sensitive contexts. Self-directed construction processes, on the part of learners, teachers and the general public, are now valued and facilitated in formal education and informally in science museums. Understanding natural science becomes a phenomenon-oriented experience with hands-on activities and culturally integrated narratives. Science education has moved from the learning of mere facts or procedures towards an understanding of science as a cultural activity. Taken together, these trends contribute to the emerging and diverse uses of historical devices in formal and informal education that are presented in this volume.

Typical science education at the end of the twentieth century was highly decontextualised. By contrast, in settings where historical scientific instruments are introduced into classrooms, the historical context is suggested, and becomes a worthy basis for discussion. In the case studies assembled in this volume, instrumentation is a form of time travel that transports characteristics that are distinctive of the era of its original making, design and use, into the present. These characteristics may encompass material, design, performative, and social facets. Instruments featured in this volume exemplify era-specific characteristics, such as eighteenth-century electrostatic devices ranging from glass rods and rabbit fur to the tremendous "Ongemen Groote Electrizeer-Machine" held at Teylers Museum, Haarlem, The Netherlands, or

© ELIZABETH CAVICCHI AND PETER HEERING, 2022 | DOI:10.1163/9789004499676_002

the early twentieth-century smoked-drum kymograph and spinthariscope. Such apparatus would not be conceived, built and used today – and yet – they are! Readers of this volume will find museum-goers shocked in activities rubbing rods with fur or marvelling at the operation of a recently-constructed replica of Van Marum's great electrostatic generator, as well as students envisioning the bodily strains borne by observers of a spinthariscope's subtle flashes. To viewers and learners today, the strange appearance and unusual materials from which these devices were made creates curiosity while revealing the materiality and design of their time of origin. In contrast to the tendency of today's technology to supplant physical materials in formal education, this volume relates how actual siphons, galvanometers, stereoscopes, anorthoscopes, and acoustic sirens are invited into the classroom, study space, museum exhibit and video portrayals.

Learners and the general public interact with physical apparatus relating to instruments of historical science in this volume's chapters. That apparatus may be: original instruments from a historical era; physical models of instruments professionally produced; replicas resulting from systematic research; materialisations of the principles underlying historical instruments; or analogue devices constructed by students. In the case where the historical originals may be fragile or pose safety hazards, chapters describe alternatives, including models, replicas, and substitutions. The actions that students undertake are diverse: viewing, in person and by video presentation; contemplation; holding and touching; manipulation and operation; using, observing, and experimenting with; examining; describing; drawing, sketching and making diagrams; discussing with others; researching; photographing; producing videos and animations; exhibiting, and serving as a tour guide for an exhibit; repairing; reconstructing; recreating; modelling; reinterpreting in creative modes; presenting at congresses; documenting and analysing in reports and theses; and using to engage other learners. Consequently, the roles of students and learners differ significantly from the roles assigned to students and learners in conventional instruction. Chapters describe learners undertaking a range of roles including: classical learners; researchers; novices being inculcated into a new field or a new perspective of science; experts preparing information on an instrument that will enable other people to understand or use it. These actions and roles set out an expansive space of possibilities for bringing students together with physical artefacts relating to historical science.

The scope and context by which these interactions come about is shaped by local specifics. A dominant mode of access to historical instruments is through museums. Through cross-fertilisation with science centres, many museums now offer activities in the context of their collections, expanding beyond the traditionally static exhibits. Examples of these experiences – each taking advantage of specific collection or site assets – are described by chapter authors who work in museums as curators, or who collaborate with a science centre or university museum. Other authors include teachers or instructors who encourage reconstructions and recreations by students to supplement their differing access to historical instruments and collections. Both active and retired, chapter authors work in diverse roles as: educators at museums, high school and university levels; museum curators and volunteers; researchers; physicists; and collection specialists.

Each volume author or collaboration presented their educational work with historical scientific instruments at a Symposium of the Scientific Instrument Commission (SIC) of the International Union of History and Philosophy of Science and Technology (IUHPST). We co-organised two sessions titled "Instruments in Education Today" at the *XXXIX SIC Symposium* held virtually in September 2020 with organisation based in London, UK. Chapters in this volume relate specifically to the contributors' personal work and expertise. The contents thus omit many areas of science, categories of historical scientific instruments, historical eras, geographical regions, educational settings, cultural communities and perspectives.

These chapters demonstrate substantial potential in using historical scientific instruments in contemporary education. As leaders in beginning to explore these potentials, we thank this volume's authors for: educational vision, resilient energy, observant descriptions, responsiveness in writing, and reflective insights. Further opportunities for bringing people together in educational encounters with historical scientific instruments have yet to be imagined, identified, acted upon and realised.

We welcome future initiatives and undertakings in this new area of science education. Alongside the challenge and fun of doing that work, we advocate for integrating into future educational effects: practices of documentation in multiple media including internet communication and presence; evaluation surveys and reports; personal and group reflection; writing and publication. In this context, the challenges, limitations and opportunities of media-supported educational processes have become apparent in the current (at the time of writing) Covid-19 crisis, which will undoubtedly make new integrative formats feasible in the future and thus offer perspectives that differ from those presented here.

We offer the experiences and projects of this volume, as encouragements to form dialogues among museum curators, educators, teachers, students, scientists and the general public. What settings and activities intrigue students and the wider public to engage with – and learn from – instruments that were used to create the knowledge that is still relevant today? How do learners' encounters with instruments develop bodily experience and conceptual knowledge that is new for them? What awarenesses do they develop about how knowledge was created and why it was (and is) considered valid? What new themes and questions might emerge, for learners, the public, educators, and museum curators and staff? To the resourceful openings of this volume's authors, what might you extend, develop and share? We invite your initiatives and participation.

Curator Jan Waling Huisman exemplifies the creativity of the volume authors, as he brings to life the heritage of the Museum of the University Groningen. Being astute about the workings and failings of original instruments allows him to apply these insights in returning originals to working order and in producing replicas. After modifying his collection's early twentieth-century smoked-drum kymograph for continual non-destructive use, he personally operated it throughout a ten-day exhibition. By making visible a historical practice, this demonstration raised public awareness of how medical research impacts on people's lives.

Whereas Huisman's expertise with and possession of instruments was critical to his programme, non-curatorial staff at the Geneva Museum of the History of Science engage the public in workshops developed by curator Stéphane Fischer. The workshop apparatus, a representation of the inclined plane of Galileo Galilei, is resilient to repeated use. When a ball is released from the ramp's top, an audience volunteer starts a stopwatch, as the staff guide invites the whole audience to observe and discuss. Based on that discussion, an audience member repositions bells along the ramp, which the descending ball strikes and rings. Audience members test their assumptions about motion by proposing and conducting multiple trial runs, yielding outcomes that surprise them while reflecting Galileo's findings, as related by the guide.

In a related way, a guide requested audience manipulation of replica electrostatic instruments during an electrical salon tour established by Peter Heering within the science centre Phänomenta Flensburg. In contrast to the decontextualised framing of this science centre's other interactives, in Heering's exhibit, a historically-trained guide invited visitors to step into eighteenth-century science both through her or his narrative, and by their personal actions, such as experiencing electrical shocks and being startled by an exploding "thunder house". Tour visitors used replicas that Heering and colleagues had constructed on the basis of their research. The value to visitors of these personal, hands-on interactions with phenomena and instruments was confirmed by interview and survey feedback.

Whereas Heering intervened in a science centre by inviting visitor interactions with replicas, Teylers Museum curator Trienke van der Spek transformed a traditional passive museum to offer interactive laboratories. Faced with the eighteenth-century electrostatic generator which is still the world's largest – but too fragile to reactivate – and endowed with spare rooms in historical décor, van der Spek commissioned an almost full-size replica to be as authentic as feasible, and operable. Staff re-enactors – and audience volunteers – hand-crank the generator during demonstrations. With its sparks crackling, awed visitors glimpse the original's electrical power. At worktables in an adjacent laboratory, visiting physics students conduct self-motivated experiments with working replicas and resilient original devices. Through this combination of viewing authentic museum artefacts in galleries, along with experiencing electrical phenomena first-hand, students are encouraged to collaborate with historical scientists, questioning similar effects and devices.

Conventionally, museum educational experiences are informal, and not coordinated with academic programmes. By contrast, chapter authors participate in a range of collaborations where museums interface with educational institutions at all levels, ranging from elementary grades to doctoral research projects. Collaborations facilitate continuity between classroom and museum. For example, as follow-up to experiments at the Teylers Museum laboratory, physics students write reports and discuss their findings in class.

Young children's curiosity is sparked by historical devices that look different and work differently from everything they see today. Historical instruments that work without electricity, including the slide rule and the gramophone, are mind-boggling to schoolchildren making their first-ever visit to the University of Rennes, in the context of a class experience with the curator Julie Priser and colleagues. In advance of each visit, Priser learns in-depth from the schoolteacher about their current studies, and makes selections that enhance classroom themes.

Schools that own and maintain historical science instruments, such as French high schools and universities having membership in the *Association de Sauvegarde et d'Étude des Instruments Scientifique et Techniques de l'Ensignement* (Association for Preserving and Studying the Scientific and Technical Instruments of Education – ASEISTE), provide the potential for youth to encounter historical scientific instruments within their own school.[1] At two ASEISTE member high schools having significant permanent collections, Françoise Khantine-Langlois, Alfonso San Miguel and Pierre Lauginie describe how high school students are tour guides during visits by: children who might

---

1 Another example of school instrument collections is discussed by Marta Rinaudo, Matteo Leone, Daniela Marocchi and Antonio Amoroso, "The Educational Role of a Scientific Museum: A Case Study", *Journal of Physics: Conference Series* 1287, 012050, (2019), available at https://iopscience.iop.org/article/10.1088/1742-6596/1287/1/012050/meta (accessed 9 Apr. 2021).

attend in the future, parents, and the general public. Sometimes a laboratory activity complements these tours; one such child participant was so enthused by the activity as to redo it at home!

A school may possess historical scientific instruments and yet be unaware of their historical significance and educational potential. Such was the case for the Maraslean Teaching Center in Athens, where late nineteenth-century laboratory instruments, crafted by top European makers, lay neglected for seventy years or more. Recently, that collection attracted the notice of physicists Panagiotis Lazos, Constantina Stefanidou, and Constantine Skordoulis. Through their collaboration with high school staff and university faculty, these instruments became the focus of an educational project. High school students and teacher-training students explored and collaborated in teams. With an authentic instrument presented as an unknown object, each team thoroughly described it; researched its history and function, sometimes contributing definitively to its identification; conceived experimental uses for it; and evaluated its condition. High school students proposed – and then carried out – reversible repairs. Teacher-training students created informal moments of teaching with the students. Working with actual physical instruments, these students built their own knowledge collectively. They experienced investigation first-hand, uncovered the history embedded in the instruments and participated directly in the continuation of that history.

While few schools are endowed with original historical artefacts comparable to those of ASEISTE members or the Maraslean Teaching Center, and school visits to museums may not always be an option, enterprising educators carry out alternative experiences. High school students researched, designed and constructed their own interpretations of instruments from historical physics as part of their work in an elective school course that their teacher created with support from Heering. The instruments made by students, such as a *camera obscura* with a lens, were then available for use and experimenting by other students in the school. Continued over time, this programme builds up a repertoire of student-constructed historical instrument analogues at each school.

Drama is an opening through which high school students use and interpret historical scientific instruments in academic courses and extracurricular programs that Flora Paparou innovates at a high school in Athens, Greece. Students act in dramas having action and mystery, that take place in their school laboratory, which is fitted with today's versions of historical apparatus. After reading Greek translations of historical science texts, Paparou's students experiment directly with the equipment. In discussing the historical and social context, students realise how their experiences corroborate with the process of science, and come to value unexpected and inconclusive findings. Taking these insights further, students and Paparou put on science theatre performances for younger audiences. As Sherlock Holmes enters the stage and the audience participates as witnesses, they turn to historical scientific apparatus for clues in solving the mystery posed by the plot.

One also encounters intriguing mysteries in the ongoing study of nature. Instrumentation is indispensable to research into behaviours that pose mysteries as to how we understand nature, such as radioactivity. With the highly sensitive instruments that they developed, Pierre and Marie Curie detected and measured radioactivity originating from quantities of elements too minute to be determined by the chemical identification methods of their day. While the Curies' original instruments were contaminated, instruments designed by the Curies and purchased at that time by physicists at the University of Rennes were never used with highly radioactive materials. These instruments survive and remain functional and safe today. Taking advantage of this extraordinary collection asset, curators Julie Priser, Dominique Bernard and colleagues created an exhibition which invites the general public into the history, discoveries, and scientific methods of twentieth-century physics, through displays, videos, and a live demonstration! Museum visitors are enthusiastic upon seeing these instruments in real use as the curators reconstruct the Curies' experiments with weakly radioactive materials. For those distant from Rennes, an online YouTube video of a 2015 demonstration features the instruments at work.

Paolo Brenni engages the public with the actual motions, sounds and visual effects of devices through online YouTube videos. These videos are Brenni's lively response to the public's desire to see something actually working, as expressed to him during his career. In collaboration with a colleague and a videographer, Brenni has produced over one hundred videos, each a few minutes long, of authentic instruments in action and exhibiting phenomena of: optics, thermology, acoustics, mechanics, electrostatics, electrodynamics, and more. These videos, which document operating devices without putting forth specific interpretations, are thus "open" for use in diverse educational contexts. Brenni takes the reader through the systematic process he has developed: selecting an instrument having a clear and "spectacular" function; establishing methods of its operation through historical and other research; evaluating the instrument's condition and suitability for reactivating; addressing safety hazards; composing a plot; filming overall and with close-ups; editing; and captioning in English and Italian.

By this method, with his expertise, Brenni brings back to life instruments that were originally used for educational demonstrations and activities at the nineteenth-century Istituto Tecnico Toscano, now preserved at Fondazione

Scienza e Tecnica in Florence, and at Liceo Paolo Sarpi in Bergamo. Historical instruments that were originally dedicated to instructional purposes, such as those figuring in Brenni's videos, fall into a category that scholarly discussion of scientific instruments has traditionally treated as peripheral, a stance that some recent studies rectify.[2]

Former teaching instruments figure prominently in the collections of schools and universities that are involved with several educational projects discussed in this volume. For example, instructional physics apparatus at the University Claude Bernard Lyon 1 (UCBL) France is restored to running order and videoed by university physics students of Khantine-Langlois and San Miguel. Brenni's stated aspiration for his video project – that it might inspire others to produce videos of historical instruments in action – is fulfilled by these students' evocative YouTube videos set to twenty-first century music soundtracks!

Historical instruments having a distinctive provenance through their past use by the nineteenth-century physicist Joseph Plateau in his revolutionary use of experimental demonstrations during lectures, make up "the cabinet of physics of Plateau" now held by Ghent University Museum (GUM) at the university where he taught. Guided by Roland Carchon and Danny Segers, a master's degree student developed the thesis to characterise instruments in this collection. Similar to Brenni's practice, the student evaluated which instruments could be modified to temporarily operate again, after minor repair and with minimal risk. The student filmed those instruments in action in reconstructions of Plateau's original demonstrations. Authenticity of these reconstruction demonstrations is enhanced by the student's research of manuscripts containing conscientiously recorded notes written by students while attending Plateau's lectures. As with Brenni's videos, this master's student's videos present the functioning of instruments that are otherwise preserved in showcases. Museum visitors can now watch Plateau's demonstrations by accessing these videos on an iPad that accompanies the displayed instruments.

---

2 Among these studies are several articles published in Marcus Granato and Marta C. Lourenço (eds.), *Scientific Instruments in the History of Science: Studies in Transfer, Use and Preservation* (Proceedings of the *XXXI Scientific Instrument Symposium*, Rio de Janeiro, Brazil, 8–14 October 2012), Museu de Astronomia e Ciencias Afins, Rio de Janeiro, 2014. See also Peter Heering and Roland Wittje (eds.), *Learning by Doing: Experiments and Instruments in the History of Science Teaching*, Steiner, Stuttgart, 2011. Other examples include: Jane Insley, "Paper, Scissors, Rock: Aspects of the Intertwined Histories of Pedagogy and Model-making", *History of Education* 46, 2 (2017), pp. 210–227; Steven C. Turner, "Changing Images of the Inclined Plane: A Case Study of a Revolution in American Science Education", *Science & Education* 21, 2 (2012), pp. 245–270.

Across this volume's chapters, historical scientific instruments provide the context for educational activities where students, or the general public, discuss, act, conduct and communicate research, including as part of a formal qualification, like a thesis. For example, while investigating GUM's single-lens microscope that resembles those of Antoni van Leeuwenhoek, a bachelor's student analysed the fluorescence lines it exhibited after X-ray irradiation. This student found that the elemental composition of its brass mount correlates with that of brass characteristic to that historical era. Electromagnetic instruments in the workroom laboratory at Teylers Museum take on a related educational function, through being operated directly by high school students who develop experimental inquiries and write reports. In a similar way, the inclined plane at the Geneva Museum of the History of Science serves as a testing ground for audience discussion and experiments with motion.

Education takes place as the interacting that goes on between learners and historical scientific instruments (including originals, replicas, reconstructions and videos) facilitates development in their thinking, discussing, acting and understanding. This close ongoing relationship between educational experiences in a museum or class, and this development of learners' capacities and autonomy, has roots in researches of Jean Piaget and Bärbel Inhelder, and others.[3] That legacy is explicitly addressed by Elizabeth Cavicchi, and by co-authors Frédérique Plantevin and Pietro Milici, as grounding for their educational philosophy in involving university students with building their own interpretive responses to historical scientific instruments.

Historical scientific instruments provide an opening to the unknown, history and direct observation for university students participating in the exploratory seminar that Cavicchi teaches at MIT's Edgerton Center. On the first day of class, the students were invited to handle several non-museum quality instruments. In relating the questions and astonishment that students voiced, and the actions that they took with these artefacts in increasingly collaborative experimenting, Cavicchi narrates development in the exploratory process. Beginning by realising that these instruments were unknown to them, the students initiated actions and discussions, through which they conceived tentative inferences about the instruments' function and context. That exploratory process is education: education by means of investigatory action and reflection

---

3　Jean Piaget and Bärbel Inhelder, *The Child's Conception of Space* (trans. by F.J. Langdon and J.L. Lunzer), W.W. Norton, New York, 1967; B. Inhelder and J. Piaget, *The Growth of Logical Thinking from Childhood to Adolescence* (trans. by A. Parsons and Stanley Milgram), Basic Books, New York, 1958; J. Piaget, *Science of Education and the Psychology of the Child* (trans. by D. Coltman), Orion Press, New York, 1970.

by learners and teachers. Readings from historical texts involve students with the thinking and actions of others before them. Through experimenting with models and reconstructions of their own making, the students form personal connections to historical instruments and analyses. That context of interactively doing history and science deepens on class visits to the MIT Museum. Appreciation, questions, and wonder arise as the students explore, and observe with, original artefacts, such as the weighty and beautifully-crafted surveying tools used by previous generations of civil engineering students.

In their teaching of future teachers of mathematics, Frédérique Plantevin and Pietro Milici developed a university course at Université de Bretagne Occidentale (UBO), Brest, France, that disrupts the unidirectional hierarchy and exclusive reliance on symbols, typical in conventional mathematical education. Many sessions of this course meet in a combination of museum and workspace, the Cabinet of Curiosity, which is equipped with historical scientific instruments (there to be handled and used) and books, as well as tools and materials for constructing one's own contraptions. In this open setting, students use, interpret, and invent mathematical instruments with historical origins. Mathematics that is presented solely through algebraic forms in most French school textbooks, takes on geometric, spatial and material relations that are new and experiential. Affixing one end of a string to a wooden rectangle's top, whose base they made to slide along a straight edge, and affixing the string's other end to a pin fixed on the paper beneath the wooden frame, they create a parabola by physical means, having a pencil point mark on the paper while travelling with the string-rectangle junction, as the rectangle slides. This instrument-mediated activity involves teacher-training students in reconstructing their mathematical understanding in new ways, thereby undergoing the changes in intellectual perspective and in building personal autonomy for mathematical thought that Piaget characterised as development. On being invited to reflect on the educational potentials of these experiences, the teacher-training students proposed a curriculum for engaging their future pupils with physical instruments, such as those they devised. As in Heering's high school project, the instruments made by these teachers remain in the Cabinet of Curiosity, enriching the working collection and available there for subsequent students.

History is extended through personal experience, and historical sources are three-dimensional in the elective course that Janet Laidla pilots at the University of Tartu, Estonia, in collaboration between the Institute of History and Archaeology and the University of Tartu Museum. As contrasted with the Brest Cabinet of Curiosity, where every implement is available for functional use, in this museum setting humanities students are initiated into professional museum practices of object handling and inspection, photography and lighting, and inventory entry. Applying these skills, each student researches a museum artefact of personal interest, self-chosen from among those Laidla specifically selected for class study. Particular to each object were surprises and ambiguities that inspired students to investigate by doing hands-on activities and background study. In making discoveries by acting on their own questions, students become committed to their own learning, and contribute to knowledge about the museum's collection. The reciprocity of learning and experience, which developed between students, objects and museum, was celebrated by opening a public exhibit where students presented their findings on posters that they designed, printed and displayed.

For these humanities students, as for the students and public audiences described in other contexts across this volume, what they experienced while engaging with historical scientific instruments was both educationally stimulating, and yet disparate from formalised education. In grappling with that disparity, Alistair Kwan identifies how scientific instruments involve learners with multiple dimensions beyond the two that Laidla references as typical of humanities instruction, including senses of touch, visceral feeling and the whole-bodily actions by which a user puts an instrument to use in operation or observation. When the whole body interacts with something, the indoors or outdoors context matters and our conceptions stretch in new ways, ranging across possibilities that are not exposed under the constraints that are conventional in educational exercises.

Into that space widened through personal and bodily experience, Kwan invites learners to notice, question and research design. Design, whether of artefacts, experience or environment, carries evidence of past and present use, struggle, culture, and practice. By investigating design, learners discern human aspects and stories that are often suppressed from view. For example, the absence of a handle on textbook portrayals of sextants confounded Kwan's student, who reacted by constructing a sextant bereft of handle. A handle's placement and form is integral with human decisions and actions germane to the instrument's use. That experientially-rooted awareness restores humanity to all knowledge and acts of knowing – a humanity that Kwan finds excised from formal education. As these activities with historical scientific instruments and their design facilitate learners' discernment of deeper humanity and function, how might education be open, by design, to facilitate the whole body of learners' experience and agency?

While the case studies gathered in this volume differ in terms of both the social sites where education takes place and the historical instruments being used, many provocative possibilities for these two aspects are not related here. Geographically, this volume's studies are limited to Western Europe and North America. Through ongoing and emergent creative collaborations among educators and museums in Asia, Africa and Latin America, new forms of educational experiences and processes engage students and the public with historical instruments by diverse, creative means. The historical instruments discussed in this volume's case studies figured in Western historical science. Educational activities that are not restricted to Western science have scope to encompass such areas as: Arab research; astronomical observations in Asia or Latin America that long predate the Renaissance, and indigenous science.

Being innovative in each local setting, the educational programmes presented in this volume have yet to attain the administrative support and educational standing that tend to be necessary for establishing and sustaining a programme over a long term. The groundwork that makes for stability and continuity in educational programmes is built through institutional, community, and educational collaborations, policies and practices. By what modes do new innovations, such as this volume's educational activities with historical instruments, become valued and integrated into the relevant organisations? What changes in educational practice and institutional structure might be entailed? Is there a role for systematic empirical surveys that go beyond the pilot studies of some volume contributions? In the example of surveys oriented around such identity characteristics as gender, age, cultural, economic, or social background, what findings of perceptions are generative for establishing educational programmes that come to be recognised as relevant by these audiences?

The diverse voices of students, learners, and the general public have yet to come into full-bodied expression in accounts of educational experiences, in this volume and elsewhere in education. How are learners encountering, investigating and understanding the historical instruments? Evidence of these experiences are most compelling when rendered in the learners' own voices and through their collaborations with historical instruments. Media technologies make it increasingly accessible for the learners and the general public to produce and communicate their experiences in reflections, reports, exhibits, photographs and videos – as documented in several chapters in this volume. In a practice of listening to, and giving weight to, the perspectives of participants, what awareness emerges for the educators who design programmes? Are the historical scientific instruments serving a science education, a historical education, or do such categorisations not make sense to participants in the educational offerings presented here?

One of many possible responses to these questions provides the cover image of this volume – the photograph of young people engaged in experiments on electrodynamics in the Lorentz Laboratory at Teylers Museum (see *Chapter 13*). The photograph clearly shows the commitment of the young people and their direct interaction with the historical instruments on the laboratory tables before them. Looking closer, we see that the young people are being addressed intellectually while at the same time, they are having fun in dealing with the historical instruments. And, in our opinion, this interrelation of the learners' minds with their whole being is also one of the central statements made by the case studies in this volume: historical instruments enable learners to have intellectual pleasure, thereby extending their natural science competence, and to simultaneously experience self-perception and self-efficacy with regard to natural science as a cultural activity. And this seems to be a particularly compelling educational goal in these times.

The editors would like to thank a number of colleagues for their support in the realisation of this volume: first of all, Giorgio Strano and Alison Morrison-Low. Giorgio Strano was enourmeously helpful in producing a detailed feedback to the first version to the authors that enabled all of them to clarify their arguments. We also thank Alison Morrison-Low for collaboration with the index and checking the overall work; her close attention improved its consistency and clarity. Special thanks go to Alison Boyle and Louise Devoy, who responded to the pandemic by transforming the SIC meeting in London into a great virtual event where a number of papers collected in this volume were presented. We thank all the volume authors and co-authors for the vision and fascination of their educational projects, and for their writings that bring about wider educational awareness and possibilities. And finally, our thanks go to the SIC community, which is a very open and welcoming group of researchers who maintain various discourses on historical scientific instruments in a lively academic tradition. We look forward to continuing these discourses together again in a classical form in 2022.

CHAPTER 1

Reading Instruments for Historical Scientific Practice

An Experiential Pedagogy for Material Culture

Alistair Kwan | ORCID: 0000-0003-3890-9650

1 Plato's Curse

Teaching with or about historical scientific instruments can be hard.[1] Students commonly respond by classifying the object or its purpose: "It's a spectroscope." "It's for measuring how much a dog salivates." "It proved that most of the atom is concentrated in a tiny nucleus." The classification instinct is strong, and only natural, for classification and correctness often dominate how science is taught. Even as curricula prescribe a focus on process and the nature of science, assessment commonly emphasises the faithful rehearsal of well-defined knowledge received from all-knowing experts. Expert knowledge includes knowledge of scientific instruments, for scientific instruments are designed for experts to use, so only experts are expected to understand them. To the rest of us, the instrument may be mysterious, even unintelligible: a steampunk contraption of brass and glass, or a sci-fi panel of buttons and dials and flashing lights. Once the object's mysterious identity has been profaned by revelatory classification, many students think that the job is done. By naming the object, students demonstrate competence in the way typically modelled by experts in lectures and on television: they give the thing a name, perform expertise through jargon, and perhaps assert an important application for the knowledge that the device is used to produce (most likely in the passive voice). Following cues in curriculum and popular media alike, our students privilege formal, codable knowledge that can be put into words, explained in lectures, and written up in journal articles, books and archives. The historical object, in such settings, is not an evidentiary source about scientific or technological

---

1 The first sketch of this essay was presented at the *XXXIV Scientific Instrument Symposium* (Turin, 7–11 September 2015), part of which was printed as Alistair Kwan, "Interpreting Tools by Imagining Their Uses", *Journal of Museum Education* 42 (2016), pp. 69–80. The ideas were further developed through presentations and conversations at CAUMAC (Canberra, 2018), ICOM – ICR/ICTOP (Auckland, 2018), and the *Knowledgeable Object Symposium* (Macquarie University, 2018), and of course through the generosity of this volume's and series' editors.

© ALISTAIR KWAN, 2022 | DOI:10.1163/9789004499676_003

history, but only a material echo of The Sacred Word, a mere handservant of the knowledge that matters.

An actual expert's knowledge of the instrument, of course, goes far beyond what books and archives contain.[2] Moreover, there are many different experts whose knowledge only partially overlaps. The expert *operator*, for example, knows how to use the instrument: how differently it behaves on humid days, how to turn the knobs without losing accuracy to a loose screw, which windows or dials to watch when, what the occasional wobble on the left signals, and how to prevent that wobble from happening. Expert operators might speak of the instrument in the passive, but they operate their instruments in dialogue, responding to the instrument, even collaborating with it. The expert *maker*, in contrast, knows that certain odd holes and notches are for registration during milling or assembly (or were actually mistakes that can be excused by calling them registration holes), that certain parts must be fitted before others, that certain adjustments are going to entail re-applying half of the lacquer. The expert *laboratory technician* knows how which consumables work best (perhaps contrary to the maker's advice), and which eyepiece or valve was actually borrowed from another instrument. Expert knowledge – from the plurality of experts that instruments may involve – is rich, multiple, thick.

But what if you aren't an expert? Chances are that none of our students are experts. For instruments from a century past, we teachers will not be experts either, at least not the experts who designed or made the instrument, or who used or maintained it in everyday work. While original context is certainly among the goals of material culture analysis, the largely practical knowledge of instruments – and tools more broadly – can be difficult to extract, and may be missed altogether as attention inclines towards names, theories and categories. This essay searches for a theoretical perspective by which to complement that formal, codable knowledge with the kinds of historical expert knowledge that the object's makers, operators, technicians used to possess. Even though that knowledge is largely practical, our journey here will be more theoretical: the goal is a bridge, or at least its footings, that will contribute in a historiographically significant way to the teaching and learning of history of science. Our starting point is something that many instruments share in common: that they were *designed* to serve a purpose, and that they were *designed* for people to use them. Because design is deliberate, it offers a way to understand an object in front of us.

---

2 See also Paolo Brenni's chapter in this volume; *infra*, pp. 34–49.

2 Design Theory and the Psychology of Our Surroundings

Many scientific instruments were designed for operation by 'typical' human operators, expressing thoughts through 'typical' human bodies.[3] Good tools 'fit' our bodies. We are good at identifying actions that tools invite us to perform: we instantly spot buttons to push, handles to grasp, levers to pull, holes to peer into. Spotting them is so natural that, when an object does not respond as expected, we may reflexively blame the object or designer without even noticing whence our expectations originated – we perceive the handles, levers, viewports, buttons, knobs or switches without any effort at all, but *how do we know what they are*?

The process of perceiving these action-components is both tricky to spot and tricky to work with, hence our struggle to foresee and correct shortcomings that undermine disabled people, for instance, by lacking good ways to think about them. Or we answer confusion with an exhortation to read the manual more carefully, not aware that something is systematically guiding operators in a different direction. Several key ideas, however, have been brought together as Design Theory, enabling them to be used both for design, and also for the analysis of spaces and objects – whether designed or not.[4]

Design theory originates in ecological psychology, an approach that treats action and perception as inseparable. In James Gibson's formulation, which has recently gained prominence in education theory, the environment presents potential for action. Objects' shapes, their locations, and their dimensions communicate possibilities like 'pushable', 'graspable', 'step-on-able'. In other words, we perceive not merely forms, but also what we can do to them. A button typically communicates pushability by standing slightly proud of the surrounding surface, being about the size of a fingertip, often with a slight depression to welcome and direct the finger's touch, and often a thin gap between it and the sleeve or faceplate that it slides into. A handle declares its graspability by being thin enough to wrap the hand, or at least fingers, around. It may be raised clear of the surface behind it so that fingers can fit into the gap; it may swell in the

---

3 Exceptions might include automated instruments, or instruments operated by computers. Space telescopes, remote-sensing satellites, self-registering data loggers and wireless-activated or animal-triggered observation stations, for example, would resist the methods outlined here. So would instruments accessed through an intermediary, such as isolated sensors engaged via a cable or a computer.

4 For a full outline of design theory and its grounding in Gibson's ecological psychology, see Klaus Krippendorf, *The Semantic Turn: A New Foundation for Design*, CRC/Taylor & Francis, Boca Raton (FL), 2006; on the emotional functions of design, see Don Norman, *The Design of Everyday Things*, Basic Books, New York (NY), 2013.

middle to nestle into the palm, or have a sequence of grooves to accommodate the fingers, or knurling that declares a better grip. A 'step-on-able' surface is horizontal and large enough to fit a foot or two; if somewhere between floor and knee level, it presents as an opportunity for climbing up or down, especially if there are several of them in a regular sequence, that is, stairs. A larger horizontal surface at knee level can read also as 'sit-on-able'. These mechanical potentials, in James Gibson's parlance, are called "affordances".[5] Affordances underpin the skeuomorphic design of computer screen interfaces that look like physical reality: on-screen buttons and slide-switches mimic their mechanical counterparts in order to communicate, without words, how to operate them. To put it another way, we perceive the world through functionality, what an Aristotelian might call *telos* or final cause. A garden path, for example, is not only the matter of its paving stones nor the long, thin form in which they are arranged, but also the line of footholds that the stones, arranged in this way, affords. Its final cause is to be walked along.

On top of affordances, there can be cultural and logical information. Two buttons arranged side-by-side are likely to have opposing outcomes: on and off, or left and right if aligned horizontally, or up and down (or forwards and backwards) if aligned vertically. A row of identical buttons is likely to activate gradations of a single action, such as for selecting the destination storey in an elevator. Colour-coding, symbols or a numerical sequence may offer further guidance – in an elevator, a star often indicates the main street connection regardless of how the storeys are numbered. Extending Gibson's purely physical affordances to embrace meanings cued through cultural norms or abstract reasoning leads to a more complex phenomenon that goes by various names; here, we will use Krippendorf's term, "affording".[6]

Getting the affordances and affordings right may well be design's most important goal. When design is 'good', no one needs instruction: the affordances and affordings do their job so well that the user effortlessly intuits what

---

5   Gibson's definition of affordance is diffuse and discursive, and was spread over many years. While not the first clear instance of the concept, a source that has come to be canonically cited is James Gibson, *The Ecological Approach to Visual Perception*, Erlbaum, Hillsdale (NJ), 1986, pp. 127–145. The understanding of 'affordance' taken in this essay is strictly ecological and perceptual, following Gibson. It does not include functionalities that do not communicate themselves within the operator's immediate perceptions, and hence differs from the broader 'what it can do – even if the operator can't tell' sense of affordance (more accurately called functions or potential) widely meant in the e-learning literature.

6   For an explanation of "affording" more pointed than Gibson's, see Krippendorf, *op. cit.* (n. 4), p. 120.

to do, and the affordances and affordings go unnoticed. If they go unnoticed, however, how can affordances and affordings be seen?

Consider the object in Figure 1.1. This object often defies efforts to classify it. The object taunts us by making every feature mockingly apparent, unoccluded by ornamentation or enclosures. Touching the shiny surface, we find that it is hard and smooth. Its shape evokes space rockets and car hood ornaments as imagined in the 1960s, or extraterrestrial robots imagined in a 1990s blend of minimalism and teuthoid biomimicry. But what is it? It does not seem to say; perhaps the best we can call it is 'sculpture'.

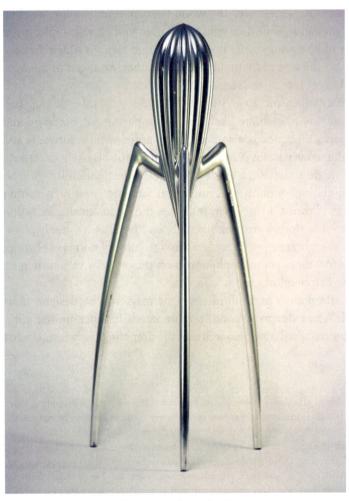

FIGURE 1.1   Philippe Starck, Juicy Salif, 1990, for Alessi
PHOTO BY JONAS FORTH, 2013, WWW.FLICKR.COM/PHOTOS/
JFORTH/8524644379, CC BY-ND 2.0

To many people, this object makes no sense until the grooved central bulb is identified as a citrus reamer. With that insight in hand, the mind swings immediately onto a different track: we imagine holding a halved lemon or orange on top, pressing down and turning, while the other hand holds one or two of the legs. We imagine the juice running down the grooves and coming together to stream off the central point below, and surmise that the space underneath is just the right size to slip a glass or small jug underneath. That is, in fact, what a buyer learns from the instruction sheet in the box.

Notice how the meaning-making shifted from considering, at a distance, an abstract, deliberate form to a concrete, intuitive imagining of bodily action, and then from bodily action to abstractly deducing the potential to slip a drinking glass underneath. Juicy Salif, a 1990 design by Philippe Starck, plays on a Plato–Aristotle complementarity between the formal and material modes of interpreting the world. The form–matter insight, not lemon juice, is what Juicy Salif really extracts.[7]

Affordances can be communicated more jarringly through poor design. Every university, for instance, seems to have a door with 'pullable' handles mounted on the 'push' side. We often discover these handles by responding to the design feature exactly as it tells us we should, regardless of the signs that command us to "PUSH". Some of us learn about this 'special character feature' of institutional 'architecture' under the squeal of a fire alarm, when pulling on the handle interrupts urgent egress: that confrontation shows how the material environment directs us far more clearly than the administrators' signage, and that it really does matter.

Such visceral engagement is exploited by Katerina Kamprani, a contemporary architect who developed a set of objects to help us "appreciate the complexity and depth of interactions with the simplest of objects around us".[8] Each of these objects is a variation of something so commonplace that its affordances and functionalities, taken for granted, go unnoticed. Each of these objects has been rendered useless by one carefully chosen change, and the

---

7 Starck is widely claimed to have said that Juicy Salif is "not meant to squeeze lemons; it is meant to start conversations". For further analysis of how Juicy Salif is interpreted, see B. Russo and A. DeMoraes, "The Lack of Usability in Design Icons: An affective case study about Juicy Salif", *DPPI03: Proceedings of the 2003 International Conference on Designing Pleasurable Products and Interfaces*, Pittsburg (PA), ACM Press 2003, pp. 146–147; P. Lloyd and D. Snelders, "What was Philippe Starck Thinking of?", *Design Studies* 24 (2003), pp. 237–253.

8 For a consideration of Kamprani's work within a broader range critiques of body–object relationships, see Simona Cosentino, Eleonora Lupo, "Dissenting Design," *Piano B. Arti e Culture Visive* 2 (2017), pp. 96–123. The gallery is on-line: Katerina Kamprani, *The Uncomfortable*, www.theuncomfortable.com (accessed 22 October 2020).

uselessness is so obvious that it draws attention to an affordance or another functional property that few of us ever really thought about. Several of the objects (Fig. 1.2) depend on handle position: a wide stockpot has both of its handles on the same side so it cannot be held level without inordinately strong forearms; a *briki* (a jug for brewing Turkish and Greek coffee) has its handle relocated directly under its spout, so hot coffee would pour onto the user's hand. These objects communicate because we naturally imagine holding the handles. Kamprani's chairs include one with a hard bulge in the middle of the seat, prompting us to imagine how it feels to sit there, and another with over-extended arms that require us to climb over them, or to be lowered in like an infant into a high chair. Kamprani's objects prompt sensory imaginations that combine proprioception and pain and annoyance and difficulty and tiredness: we imagine bodily discomfort.

FIGURE 1.2 Some of Kamprani's uncomfortable objects
PHOTOS BY SIMON BERRY, *BAD DESIGN*, WWW.FLICKR.COM/
PHOTOS/COLALIFE/ALBUMS/72157645617459712, CC BY-SA 2.0

How might these ideas be applied in the classroom? More to the point, to what new understandings of historical scientific instrumentation could those applications lead?

3   Didactic Experiences

Many art educators begin class with a simple question: "What do you notice?"[9] That question immediately shifts the emphasis off 'scholarly' categories (period, style, medium, genre, artist) to personal perceptions. Asked what they notice, learners take numerous directions, responding, for example, to material, shape, similarities, evocations, textures, proportions, size, heft. They might notice motifs that support a stylistic reading, or geometries for a structural reading. They might notice knobs or handles or other affordance-related features, leading towards a reading in terms of handling.

Sextants and octants, for example, offer a puzzling mixture of see-through circles and what appears to be an eyepiece (Figs. 1.3–1.5). There is a hinged arm whose free end slides across an angle scale. And there is sometimes – but not always – something on the back, or underside, that might look like a handle. Many people find that the arrangement makes no sense. The first few times I saw a sextant, they made little sense to me either, regardless of the many diagrams and historical mentions I had previously read. The handle was especially confounding: explanatory diagrams often did not show a handle, and many sextant handles do not look like the handles that most of us are familiar with. On some instruments, the handle looks more like a mounting bracket or a counterweight, or a decorative element yet to be carved with a coat of arms, or even just a projection to hold the sextant secure in its case. In older instruments, the handle is part of the frame, and looks like a mere strut. A student once made a sextant for a class that I taught,[10] and did not include a handle at all for there were none in the explanations or specimen photographs that he had consulted: the source documents were all about angles and

---

9   For a recent discussion of this mode of leading interpretation, see Lisa Schneier, "Give Them the Butterflies", in Mary Kay Delaney, Susan Jean Mayer (eds.), *In Search of Wonderful Ideas: Critical Explorations in Teacher Education*, Teachers College Press, New York, 2021, pp. 75–76.

10   The first iteration of this course – directed primarily towards establishing historical scientific instruments as primary sources in the history of science – is described in Alistair Kwan, "Determining historical practises through critical replication: A classroom trial", *Rittenhouse* 22 (2010), pp. 132–151.

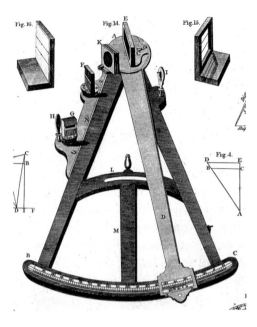

FIGURE 1.3
Eighteenth-century style octant without an obvious handle
WILLIAM HENRY HALL (ED.), *THE NEW ROYAL ENCYCLOPAEDIA*, C. COOKE, LONDON, 1791, PL. "ELEMENTS OF NAVIGATION", FIG. 14

FIGURE 1.4
Sextant and its use in a twentieth-century diagram emphasising the angle measurement
CLYDE FISHER, JOHN HIRAM GEROULD AND JAMES PLUMMER POOLE, *THE MARVELS AND MYSTERIES OF SCIENCE*, W.H. WISE & CO., NEW YORK, 1941, P. 128, FIG. 10

READING INSTRUMENTS FOR HISTORICAL SCIENTIFIC PRACTICE 23

FIGURE 1.5   Sextant and its use in a late nineteenth-century depiction
A. GUILLEMIN, *EL MUNDO FÍSICO*, MONTANER Y
SIMÓN, BARCELONA, 1883, V. 2, P. 208, FIG. 213

optics, and sextants he saw in museum displays and photographs were oriented with the handle towards the floor or back wall because the eye-catching, skill-communicating, wonder-inspiring side carries the engravings and optics. When sextants and octants are shown in use, the handle tends to be oriented away from the viewer, and covered by the operator's hand.

The handle, of course, *is* important. One hand goes on the handle, so operators can align their eyes with the eyepiece, and have the other hand free to swing the instrument's arm. When the instrument is used correctly, the handle is definitely a handle, and it is in just the right place. This insight can be developed by extending the analysis from a hand–handle connection to a whole-body interpretation that brings hands and eye into play along with the arms, neck, head and torso between them. That whole-body, whole-instrument interpretation can be prompted by questions such as the following:

- Where would your arm fit?
- Would you sit or stand – would it be different on land, versus a listing ship?
- What if you held it differently – maybe it's upside-down or sideways?
- For how long could you work like that? How would your body feel?

In my classes, the embodied approach shifted an astrolabe from being perceived as a geometric projection and decorative fretwork to a purposefully heavy object that hangs on a finger or thumb or hook, while you sight stars or spires along its alidade. This shift is doubly meaningful, for it shows that there are in fact *two* useful ways to hold astrolabes: in an edge-on orientation for measuring, and a flat-facing orientation for calculating and admiring. Astrolabes hence switch automatically between two modes according to their connection with the operator's body – a precedent for what today's gadget industry calls "context-aware functionality."

An embodied approach can likewise shift an observing couch from a luxurious venue where gentleman-astronomers lie back to ponder the mysteries of the heavens, to practical relief from the backbreaking, neck-crunching, knee-grinding labour of timing meridian transits without one (Fig. 1.6). The long telescope control rods can seem unnecessary or inconveniently located for an observer standing up, but make perfect sense for an observer restricted to a couch. On the other hand, sometimes there is no good space for a couch. An optical elbow – a reflector (usually a prism) mounted so the eyepiece can attach perpendicularly to the telescope's axis – saves the astronomer's neck. At other times, there is no aid at all: the astronomer is simply uncomfortable. What might that let our students inquire about the nature of the work? Perhaps the measurements were confined to short periods that the astronomer could endure. Perhaps several observers took shifts to share the burden. Perhaps discomfort was a trade-off to minimise the carriage of equipment to a remote or dangerous field site (Fig. 1.7). Or perhaps astronomers are a bit tougher and more hands-on than those stereotyped gentleman observers who lie back beneath the starry sky to dream of queen and cosmos.

In the same way, a spinthariscope (Fig. 1.8) shifts from being a philosophical amusement to a demand for long, careful concentration, a good posture to avoid a stiff back or neck, and a relaxed face (both eyes open, perhaps one of them covered with a black eyepatch) to avoid cramps. Ernest Rutherford's nuclear model of the atom shifts from originating in a graph of scattering distributions to a careful process of staring fixedly into that tiny lens in front of the phosphorescent screen for a whole minute or two before taking a break, and all that only after thirty minutes of waiting for the eye to adapt to the darkened laboratory so that the faint flashes could be seen at all. What did

FIGURE 1.6   Equatorial telescope at Greenwich
LEISURE HOUR 525, JAN. 10, 1862, P. 40

the laboratory spinthariscope even look like? Research articles generally show only a schematic (Fig. 1.8, bottom left), not the actual object in use.[11]

As these cases suggest, an embodied interpretation process, developed through design theory, has potential to lead learners towards readings radically different from the formalist, stylistic, mathematical and theoretical readings

---

11   H. Geiger and E. Marsden, "On a Diffuse Reflection of the α-Particles", *Proceedings of the Royal Society A* 82 (1909), pp. 495–500: p. 496.

FIGURE 1.7　Zenith telescope in the field
CANADA DEPT. OF MINES AND TECHNICAL SURVEYS /
LIBRARY AND ARCHIVES CANADA / PA-019667

more usual in science and history of science, and can fill out the "would have" statements that can otherwise want for explicit justification: a plausible justification for "they would have" may be as simple as, "Let's all try standing like that for ten minutes to see how it feels", given a classroom encultured to work with such ways of knowing. Interpretations could be expressed through models or mannequins or drawings in which human figures communicate the relationships between instrument and operator and environment, thus

READING INSTRUMENTS FOR HISTORICAL SCIENTIFIC PRACTICE   27

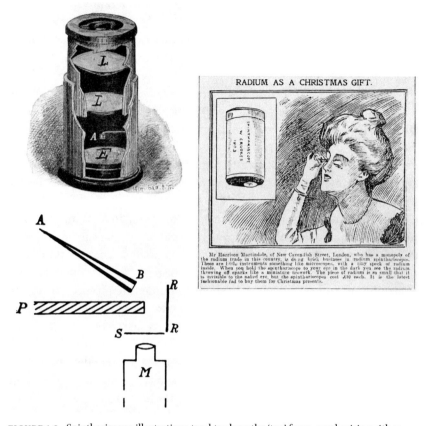

FIGURE 1.8 Spinthariscope illustrations tend to show the 'toy' form, emphasising either theory (*top left*) or amusement (*right*). As research documentation is typically schematic (*bottom left*), the screen's form, affordances and use must be deduced from extant objects and descriptions.
RESPECTIVELY FROM: *SCIENTIFIC AMERICAN* 91 (1904), P. 428; *DUNDEE EVENING POST*, 10 DECEMBER 1903, © THE BRITISH LIBRARY BOARD, USED WITH PERMISSION VIA THE BRITISH NEWSPAPER ARCHIVE; H. GEIGER AND E. MARSDEN, "ON A DIFFUSE REFLECTION OF THE α-PARTICLES", *PROCEEDINGS OF THE ROYAL SOCIETY A* 82 (1909), P. 496

complementing ideas that can be expressed naturally in text. Such discussion would centre on empirical practice, offering a perspective on history of science, technology and medicine that complements the more usual attention to theoretical concepts by adding different kinds of knowing, struggle, patience, cost, technical complication and expertise.

4 Design Analysis, Material Culture, and Historiography

Material culture theories have tended to emphasise the visible content of objects. That tendency is understandable: epistemic culture, indeed human nature, skews heavily towards the visible. As Aristotle noted long ago, we learn through sight above all other senses,[12] a preference echoed in modern metaphor: we understand by 'seeing' and 'getting the picture', we gain 'clarity' when 'illuminated' by 'lucid' or 'transparent' explanations that 'elucidate' or 'shine light upon' the 'opaque', the 'obscure', the 'darkness', hence 'lifting the veil', or 'clearing the fog'. Subtlety and context are 'perspective' and, in the pretentious patois of government and business, bias and spin-doctoring are 'optics'. Scarce are metaphors that connect scientific knowing with other senses, though there are a few associated with haptic interaction: 'to grasp', 'to fathom', and some intuitive and emotional knowing through sensations in the throat, stomach, and limbs. That same visual tendency occurs, naturally, in the canonical studies underlying material culture method. For example, E. McClung Fleming's court cupboard study[13] (which established the Winterthur protocol) relates an object's visible features to cultural context, but underplays the actions of opening the cupboard doors and putting things inside it. Craig Gilborn's Coke-bottle study[14] similarly focuses on the visible, even though much of Coca-Cola's advertising evokes the haptic – cold curved glass wet with condensation, held in the hand and pressed against the lips – and includes the auditory – the iconic sound of gas escaping from a freshly opened bottle. Jules Prown's teapot study[15] concentrates mostly on visible shape, and from here speculates about how the object *might* feel, in order to deduce the functions of the teapot's belly, handle, lid and spout. Like Gilborn's study, Grant McCracken's study of clothing[16] brings us from high culture to popular culture, while cautioning us about the

---

12   Aristotle, *Metaphysics* (trans. by Hugh Tredennick), "Loeb Classical Library 271", Harvard University Press, Cambridge (MA), 1933, 2 vols.: v. 1, pp. 2–3, (1.1–2).

13   E. McClung Fleming, "Artifact Study: A Proposed Model", *Winterthur Portfolio* 9 (1974), pp. 153–173. The Winterthur protocol can be thwarted by the strong tendency towards visible and codable knowledge: see the challenges documented (some of them only tacitly) in Katharine Anderson, Mélanie Frappier, Elizabeth Neswald and Henry Trim, "Reading Instruments: Objects, Texts and Museums," *Science and Education* 22 (2013), pp. 1167–1189.

14   Craig Gilborn, "Pop Pedagogy: Looking at the Coke Bottle", *Museum News* 47 (1968), pp. 12–18.

15   Jules Prown, "Mind in Matter: An Introduction to Material Culture Theory and Method", *Winterthur Portfolio* 17 (1982), pp. 1–19.

16   G. McCracken, "Clothing as Language: An Object Lesson in the Study of Expressive Properties in Material Culture", in Reynolds et al. (eds.), *Material Anthropology*, University Press of America, Lanham (MD), 1985, pp. 103–128.

language of visible matter: conservative codes are relatively readable because they follows stiff rules, but innovative codes, especially the fashion codes of subcultural rebellions, can be very hard for an outsider to make sense of. When looking back at past fashions (including fashions that shaped scientific instruments), we can be misled by the simple reason of having lost the language or code of an otherwise little-documented subculture (for example, that of laboratory scientists or Arts and Crafts machinists), applying today's conventions or text-documented norms with no easy way to detect their inapplicability.

Material culture readings rely also on connoisseurship and perceptiveness. Scientific instrument connoisseurship, like every other kind, comes but slowly, and few have the opportunity to develop it. As for perceptiveness, we may fear for its continuing diminution as society, and in particular formal schooling, leans persistently towards knowledge that can be handled by computers and delivered through words and simple diagrams alone. Design analysis offers a way to bring perceptiveness back to the fore, echoing and extending the descriptive phases of Prown's and Fleming's approaches, and making them accessible to novice learners by embracing more senses and the kinds of non-visual knowledge that the full range of senses provides access to.

By expanding description from visual and geometric to embrace the haptic and proprioceptive, design analysis can broaden material culture readings to shift the historiographic positioning of historical instruments in (at least) three important ways.

First, instruments become evidence. Rather than slotting available objects into text-driven narratives, learners are called to read the instruments directly for themselves, to find and construct their own meanings. Because practical issues such as bodily comfort and specific skill needs are documented only very lightly in the textual record, there is little opportunity or temptation to look up what someone else claimed about it: the instrument may be the only extant source at hand, or the only extant source at all. The instrument becomes a potential source of tacit knowledge.[17] Tacit knowledge is essential to a comprehensive understanding: as Michael Polanyi put it, in his consideration of how perception drives knowing, "A wholly explicit knowledge is unthinkable."[18] The knowledge that I am after here, though, is made tacit through more than not being said: it is not even sayable. There are no words for it, no notation by which it is systematically expressed. Such non-codable knowledge matters,

---

17   See Michael Collins, *Tacit and Explicit Knowledge*, University of Chicago Press, Chicago (IL), 2010.

18   Michael Polanyi, "The Logic of Tacit Inference," *Philosophy* 41 (1966), pp. 1–18: p. 7.

and a surprising amount can be experienced by approaching the object in a non-linguistic way.

Having construed the object as its own evidence for itself, learners have good reason to check catalogues of other collections for corroborating specimens or counterexamples. My own students found this process frustrating, as they discovered that photographs nearly always showed instruments from a standardised view (e.g. very few museum astrolabe catalogue entries show the astrolabes' sights), and with an overall emphasis on pristine or restored exemplars that show no evidence of wear or adaptation or usage context. That frustration was actually helpful: it showed them how documentation and evidence survival are often determined by the questions being asked at the time of collection, and by cultural value systems that prescribe what and whose evidence counts as 'significant'. While these biases might be considered blinkered, they can be recast as the primary evidence of collection formation processes and of historiographic considerations that learners commonly find too abstract or irrelevant, but are here made meaningful by direct engagement.[19]

Positioning instruments as evidence can also help students to understand texts. I once asked a class to tell me about the scarificators that they had been reading about in the context of therapeutic bloodletting. After a few minutes of hearing their thoughts, I opened a box. As they beheld the scarificator inside, I asked, "Is this what you were imagining?" Not one of them had imagined anything like it. I slowly cocked the spring and pressed the release. Most jumped when they heard the mechanism snap; the blades moved so quickly that half of them didn't see it happen. They hadn't imagined this, either. They saw how clumsily I held the device, and none wanted to try it themselves, fearful of slipping. "Does anyone want to volunteer an arm to see how cleanly it cuts?" They recoiled. Visceral imagination had, in only a few minutes of non-codable knowledge, transfigured what they had read, much as Polanyi declared: "... what the pupil must discover by an effort of his own is something we could not tell him. And he knows it in turn but cannot tell it".[20]

A second historiographic shift is that scientific practice becomes bodily practice. As much as curricula exhort the teaching of science as process, and as an interplay between conceptualisation and evidence and socio-cultural forces, science textbooks, histories of science, and university science education all hold experimentation down in a supporting role. Experiments are

---

19   Analogous problems occur also in the history of music, where the materiality of instruments and their relationships with performers' bodies is little-studied: see Nicholas Cook, *Beyond the Score: Music as Performance*, Oxford University Press, New York, 2013.

20   Polanyi, *op. cit.* (n. 18), p. 5.

typically mentioned as accidents providing serendipitous discovery (for example, the Stern–Gerlach experiment, William Crookes' invention of the spinthariscope), confirmations of an idea (for example, Galileo dropping balls off the Leaning Tower of Pisa, Arthur Eddington measuring the deflection of starlight), or as an ingenious technique (for example, Fizeau's and Foucault's methods for measuring the speed of light). As with the sextant mentioned earlier, however, we seldom read much about the practicalities of actually performing these experiments; they are typically described only in passing. That streamlined portrayal is echoed in science classrooms – witness, for example, the mass-produced torsion balances for measuring Coulomb's Law. While history of science can approach replication with a clearly historical purpose, overall it finds practical science elusive – what evidence does *doing* leave behind? Design analysis offers a way to bring some of the experimental process and tacit practical knowledge back into play, and to interrogate what is meant by words such as "skill" and "technique". In particular, design analysis gives opportunity to consider how a good deal of tacit knowledge comes through difficulty and discomfort. An operator's discomfort may drive improvement towards fluent action (that is, critical practice to hone skills), or alternatively to stoicism and stamina. As music, dance, and trades educators are well aware, however, uncritical repetition can also lead to the learning of bad habits.

Within the practical-science culture of doing and experiencing there lies a third shift: a profound humanisation of scientific instruments. Design analysis expands the cultural scope in which historic instrumentation can be contextualised. Adding to the practical culture of doing, and the visual culture of appreciation, and the intellectual culture of natural and mathematical learning, design analysis takes us to the individual human element of bodily engagement, of visceral experience, and of that ancient question of how our senses and other bodily capacities restrict – or determine – human access to Nature's deep truths. Do we really perceive truth, whether directly (recall Plato's cave and the Renaissance Pyrrhonists) or through mediating instrumentation (early opposition to the telescope)? Can we truly share observations (hence the solar microscope and self-registering instrumentation)? Can others replicate our results (hence the personal equation for astronomical timings)? What limitations in the mind–body–instrument–nature complex can be addressed through design? Design directly links individual humans with the knowledge that those humans comprehend and produce. Both the body-imagination inquiries described above, and historical replication methods, can thus be given clearer purpose – whether didactic or historiographic – through explicit design analysis.

5 Design Analysis and the Museum in the Curriculum and Pedagogy

To absorb and rehearse pre-processed information is not enough: as John Dewey argued, learners must *experience* knowledge, and that experience must be more than the kind that pads a curriculum vitae, but an intellectually engaging kind, an experience that transforms us.[21] Experience features strongly in early childhood education, where curriculum involves all of the senses and integrates psychomotor and affective learning with the cognitive. As children age, the curricular balance shifts towards the cognitive alone, towards words and symbols and abstracts. We welcome abstraction as intellectual maturation, but may not notice what broad swathes of knowledge are excluded by abandoning the emotional and sensory. The knowledge lost is not simply 'it makes my fingers hurt', but a far more consequential loss in the range and depth of cognitive engagement: higher education exhibits a damning comfort with the lowest levels of Benjamin Bloom's taxonomy, far short of the critical thinking and citizenship that curricula and institutional vision statements widely champion.[22]

An embodied analysis of design can help to restore what Edward Reed, following Dewey, calls "the information – termed *ecological* – that all human beings acquire from their environment by looking, listening, feeling, sniffing, and tasting – the information, in other words, that allows us *to experience things for ourselves*."[23] It matters because, as Dewey put it, "We never educate directly, but indirectly by means of the environment. Whether we permit chance environments to do the work, or whether we design environments

---

21 Dewey distinguishes, throughout his works, between sensory experiences that are merely 'had', and reflective experiences that are 'known'. See, for example, in *The Quest for Certainty*, in Jo Ann Boydston (ed.), *The Collected Works of John Dewey: The Later Works, 1925–1953*, Southern Illinois University Press, Carbondale (IL), 1981, v. 4, p. 194. On applying Dewey's experientialism in museums, see Ted Ansbacher, "John Dewey's *Experience and Education*: Lessons for Museums," *Curator* 41 (1998), pp. 36–50; on the role that museums and museum education play in Dewey's theory, see George E. Hein, "John Dewey and Museum Education," *Curator* 47 (2004) pp. 413–427.

22 See e.g. A.J. Swart, "Evaluation of Final Examination Papers in Engineering: A Case Study Using Bloom's Taxonomy", *IEEE Transactions on Education*, 53 (2010), pp. 257–264; Alison Cullinane and Maeve Liston, "Review of the Leaving Certificate Biology Examination Papers (1999–2008) Using Bloom's Taxonomy", *Irish Educational Studies*, 35 (2016), pp. 249–267.

23 Edward Reed, *The Necessity of Experience*, Yale University Press, New Haven (CT), 1996, pp. 1–2. Italics in original.

for the purpose makes a great difference."[24] Obviously, designing the learning environment entails attention to multiple sensory factors, and sensory factors must, by necessity, feature in any complete understanding instruments made for sensory use. Reed argues for more: ecological information is essential to realising individual and communal human potential, for it connects us with our contexts, freedoms and constraints. We need ecological information to tackle the age-old question, "How should I live?" Historical scientific instruments, with their close ties to how we understand and conceptualise the world in which we exercise our free agency and communal contribution, may be a good entry-point for knowledge-experiences of especially worthwhile kinds.

Design analysis will not work for all kinds of instruments, nor for all kinds of science. But, like palaeography and textual criticism, it does offer a starting-point for direct historical inquiry. It is not blocked by mathematical anxiety, bad experiences with cursive handwriting, nor by not knowing Arabic, Chinese, Sanskrit and Latin. It does not exclude cultures whose knowledge is not written. Design analysis offers museums and historical instrumentation collections a special educational role that textbooks, lectures, and lecture slides cannot cover.

---

24   John Dewey, *Democracy and Education*, Free Press, New York, 1997, pp. 18–19. For further interpretation, notably in the distinction between a crafted learning environment and mere surroundings, see David T. Hansen, "Dewey's Conception of an Environment for Teaching and Learning," *Curriculum Inquiry* 32 (2002), pp. 267–280.

CHAPTER 2

Filming Nineteenth Century Physics Demonstrations with Historical Instruments

Paolo Brenni | ORCID: 0000-0002-0165-3033

1 Introduction

In the past few years,[1] with two colleagues, I have had the opportunity to make a series of videos illustrating the use and working of historical scientific instruments, which were very common in physics teaching collections of nineteenth and early twentieth centuries. The videos were done in two different collections. Most of them were realised with the instruments of the collection of the Fondazione Scienza e Tecnica in Florence.[2] This institution preserves the very large historical collection of instruments of the former Istituto Tecnico Toscano (later Istituto Galileo Galilei), which was founded in 1850 and became one of the most important Italian technical schools. Another sixteen videos were realised in the collection of the Paolo Sarpi High School of Bergamo.[3] To our surprise these videos, which are freely visible on YouTube, are very popular reaching hundreds of thousands of virtual visitors, which is quite impressive considering their specialised character.

In this paper I will illustrate why we decided to produce these videos, and how the instruments were selected and filmed. Furthermore, I will analyse the possible conservation problems associated with the fact of using working

---

1 The first and shorter version of this article was presented at the *XXXI Scientific Instrument Symposium* held in Rio de Janeiro, Brazil, 8–14 October 2012.

2 See: www.youtube.com/user/florencefst/playlists (accessed 10 Sept. 2021). About the physics collection of the Fondazione Scienza e Tecnica see: Paolo Brenni, *Gli strumenti del Gabinetto di Fisica dell'Istituto Tecnico Toscano, I Acustica*, Provincia di Firenze, Firenze, 1986; Paolo Brenni, *Gli strumenti del Gabinetto di Fisica dell'Istituto Tecnico Toscano, II Ottica*, Istituto e Museo di Storia della Scienza – Giunti, Firenze, 1995; Paolo Brenni, *Gli strumenti del Gabinetto di Fisica dell'Istituto Tecnico Toscano, Elettricità e Magnetismo*, Fondazione Scienza e Tecnica – Le Lettere, Firenze, 2000, and Paolo Brenni, *Il Gabinetto di Fisica dell'Istituto Tecnico Toscano, Guida alla visita*, Polistampa, Firenze, 2009.

3 See: www.museovirtualesarpi.it/progetto.html (accessed 10 Sept. 2021). About the physics collection of the Liceo Paolo Sarpi see also: Laura Perani Serra, *Gli strumenti del gabinetto di fisica del Liceo Classico Paolo Sarpi di Bergamo*, Associazione ex alunni Liceo Paolo Sarpi, Bergamo, 2009. I would like to thank Laura Perani Serra, whose efforts and collaboration were essential for realising the videos in Bergamo.

© PAOLO BRENNI, 2022 | DOI:10.1163/9789004499676_004

FILMING NINETEENTH CENTURY PHYSICS DEMONSTRATIONS 35

FIGURE 2.1 Anna Giatti is ready to show Chladni's figures while the video maker Antonio Chiavacci is filming.

historical apparatus, which generally repose in their cupboards. Finally, I will explain the utility of such videos and how and where they were (and are) used.

Over the last few years we have produced about 110 videos. These results would have been impossible without the enthusiastic and indispensable collaboration of my long-time colleague Anna Giatti. Together, we studied the experiments, we performed them several times, we prepared the set for filming the demonstration, and we wrote the short explanatory texts of the videos. Antonio Chiavacci and his Hastavideo company is the professional video maker who shot the films and took care of the necessary post-production (Fig. 2.1). Thanks to his experience and skill, it was possible to produce videos

2 Why Is It Useful to Use Historical Working Instruments for Reproducing Classical Experiments and Demonstrations?

In my many years spent in museums and collections, preserving and studying historical scientific instruments, there is a series of similar questions that visitors asked me hundreds of times: Can we see instrument in operation? Why are these models not working? Could you show us how it works? etc. Many visitors are accustomed to science centres, to interactive museums where artefacts are operational and can be touched and manipulated. Furthermore, while the same visitors do not feel the need to drink out of a Renaissance chalice or to wear an eighteenth-century tricorn to appreciate them, on the other hand they want to see ticking clocks, sparking electrical machines and moving mechanical models. But many collections do not have the opportunity of showing working models, nor do they have replicas of instruments which can be operated. This is especially the case for collections that originally, in the more-or-less recent past, were used in teaching or research institutions such as schools, universities, physics cabinets, and observatories. Non-operational historical instruments, machines or models resting in their cupboards often generate a sense of frustration in the visitors, who consider these silent historical artefacts to be dead and not very attractive. It is well known that a museum room displaying ticking clocks with their slowly moving pendulums attracts the curiosity and the interest of the visitors much more than the same room with the same, but silent, clocks. People do like to see puffing steam engines, sparking electrical machines and moving orreries. Furthermore, it is true that many scientific and technical objects are not self-evident and their function is very difficult to explain without showing them working. But obviously it is impossible to continuously (or periodically) operate historical instruments or machines. Such an approach would not only require very complex and expensive logistics (the need of trained personnel, the necessity of expensive maintenance, the respect of strict safety rules) but also it would be incompatible with the most basic requirements necessary to preserve ancient artefacts. Sometimes instead of manipulating and filming original, unique or very delicate instruments it is possible to make almost perfect working replicas of them. But high-quality replicas can be very expensive and only a few of them can be made. For example, in recent times Wolfgang Engels carefully replicated a few very important instruments such as the large electrical machine made by John Cuthbertson for Martin van Marum and the pneumatic pumps of Jacob Leupold and Jan van Musschenbroek, which are used for repeating some historical experiments.[4] There is also the possibility of showing a working instrument virtually, thanks to high-quality computer animations. Such a system can be useful for virtually disassembling and exploring a technical artefact as well as for illustrating how it works. It can be an excellent didactic tool, but will not satisfy the audience's desire of admiring a real working object.

Today, many nineteenth- or early twentieth-century scientific instruments survive not only in science museums but also in the collections of schools, high schools, technical institutes, polytechnics and universities. Generally, these instruments were used until the first decades of the last century and then they were slowly abandoned because they became useless for teaching modern physics and not adequate anymore for the new didactic approaches. In a few decades, these instruments became obsolete without becoming interesting from a historical point of view. Only from the late 1970s, historians of science and technology began to be interested in these artefacts, which subsequently were slowly re-discovered and studied. Physics teaching had changed radically since the era when the didactic instruments were made. Today most of the teachers and professors do not know how to use the instruments of classical physics, and, in addition, they do not even know what purpose these instruments served! So, with very few exceptions, these instruments were never filmed in their original circumstances and, because they have been museum pieces for a long time, very few people have seen them "in action".

Finally, all the above-mentioned considerations, and the fact that we had at our disposal a large number of restored and perfectly working nineteenth- and early twentieth-century demonstration instruments, motivated us to undertake the making of a series of videos showing their working, function and use. We imagined that these videos could have a potential public audience.

---

4 Commenting on his work with the electrostatic machine and on the pneumatic pumps and accessory (some of them were reconstructed because lost), Wolfgang Engels said: "These lost experiments, which originally belonged to the pumps as original parts, have been brought back to life. Now they can be demonstrated again true to the original, i.e. including all their historical imperfections and mistakes as we have discovered. For reasons of conservation and museology, the three original instruments are not ready for demonstration and, in addition, are not completely preserved. Apart from a faithful replica, there is no other way to experience what Napoleon, Augustus the Strong of Saxony or Landgrave Charles of Hesse-Kassel may have been shown" (personal communication).

3 Problems and Hazards in Using Historical Instruments

The use of historical instruments certainly involves a series of hazards for the operators. Today's safety rules concerning scientific apparatus are in my opinion far too strict, but certainly until fifty years ago, or even later, the measures for safeguarding users were almost non-existent. Nineteenth- or early twentieth-century electrical apparatus are often poorly insulated, and they sometimes require mercury for interrupters, commutators and contacts. Mercury, which is today banned from teaching laboratories, was formerly used in significant quantities in vacuum pumps and other ancillary apparatus (manometers, gauges, etc.).

Old boilers can be dangerous, while certain devices, like batteries and cells, require the use of strong acids and poisonous chemicals (such as lead and chromium dichromate). The use of open flames produced by petrol, oil or gas was common in several types of lamps, heaters and other devices. Not only were X-ray tubes not shielded and thus diffused a lot of radiation, but also different discharge tubes (such as the Geissler's, Hittorf's, Crookes' and Braun's examples) also produced a certain amount of X-rays.

For these reasons, in our age, which is highly concerned (and I would say sometimes obsessed) with safety, many apparatuses made in the past (sometime only a few decades ago) are considered to be far too dangerous and therefore cannot be used in a classroom anymore. On the other hand, many old instruments can be easily damaged if they are carelessly manipulated. Glass tubes and elements are easily broken, rubber and hard rubber become fragile, old soldering and joints are not able to withstand high pressure anymore, and so on. Surfaces can be damaged and lacquered brass is corroded by the normal moisture of our hands: that is why the use of gloves in working with these instruments is essential. Considering that we have to minimise the risks of damaging historical artefacts, one can keep in mind a few important recommendations. For example, for operating steam engines models, it is better to use compressed air instead of steam. That will avoid the condensation of water vapour in certain parts of the model. Certainly, when we film experiments and demonstrations, all these instruments had to be manipulated with care and only by people who perfectly know the instruments, their technical characteristics, their history (restorations, modifications, alterations, and so on) and how to operate with them.[5]

---

5 The patient work of restoration made by the author together with Anna Giatti (one of the few officially qualified restorer of scientific instrument in Italy) on the collection of the Fondazione Scienza e Tecnica allowed us to acquire "forensic" knowledge of the instruments.

FILMING NINETEENTH CENTURY PHYSICS DEMONSTRATIONS 39

4 How to Understand Nineteenth-Century Demonstrations in Order to Repeat Them

Nineteenth-century physics treatises, textbooks and sometimes trade catalogues – which were our main sources of information and inspiration –, carefully describe experiments and lecture demonstrations, as well as the instruments necessary to perform them (Figs. 2.2 and 2.3).[6] But normally, these texts do not mention the tricks, the knack, and all the only apparently secondary details which nevertheless are essential for assuring the complete success of an experiment. For these reasons it is necessary to acquire a tacit knowledge, which is not important for understanding the physics laws and principles

---

This knowledge is indispensable in order to safely manipulate them as well as to assure the best condition for their preservation.

6   Here is a list of some of the most important physics treatises which we used for understanding and performing our demonstrations: Augustin Boutan, Joseph Charles d'Almeida, *Cours élémentaire de physique précédé de notions de mécanique et suivi de problèmes*, Dunod, Paris, 1862; Henri Buignet, *Manipulations de physique*, J.B. Baillière, Paris, 1877; Charles Chevalier, Julien Fau, *Nouveau manuel complet du physicien-préparateur ou Description d'un cabinet de physique*, Rortet, Paris, 1853; Reginald S. Clay, *Treatise on Practical Light*, Macmillan and Co., London, 1911; Alphonse Daguin, *Traité de physique élémentaire théorique et expérimentale avec application à la météorologie et aux arts industriels*, Ch. Delagrave – P. Privat, Paris – Toulouse, 1878–1879; Ch. Drion, M. Fernet, *Traité de physique élémentaire*, Victor Masson & Fils, Paris, 1861 (and other editions); Joseph Frick, *Physikalische Technik oder Anleitung zu Experimentalvorträgen sowie zur Selbstherstellung einfacher Demonstrationsapparate*, Friedrich Vieweg und Sohn, Braunschweig, 1904–1909 (and other editions); Adolphe Ganot, *Cours de physique purement expérimentale et sans mathématiques*, Chez l'Auteur, Paris, 1878; Alphonse Ganot, *Traité de physique élémentaire*, Hachette, Paris, 1894 (and various other editions); Amédée Guillemin, *Les phénomènes de la physique*, L. Hachette et Cie, Paris, 1868; Jules Jamin, *Cours de physique de l'école polytechnique*, Gautihers-Villars et fils, Paris, 1886–1991 (and other editions); E. Lommel, *Lehrbuch der Experimentalphysik*, Johann Ambrosius Barth, Leipzig, 1895; John Henry Pepper, *Cyclopaedic Science Simplified*, Frederick Warne and Co., London, 1877; John Henry Pepper, *The Boy's Playbook of Science*, G. Routledge and Sons, London, 1881; Johann Heinrich Jacob Müller, Claude Servais Mathias Pouillet, *Müller-Pouiller's Lehrbuch der Physik un Meteorologie*, Vieweg und Sohn, 1886 (and other editions); Augustin Privat-Deschanel, *Traité élémentaire de physique*, Paris, Hachette, Braunschweig, 1869; Antonio Roiti, *Elementi di fisica*, Firenze, Le Monnier, 1891–1894; Jules Salleron, *Notice sur les instruments de précision*, Chez l'Auteur, Paris, 1864; Wilhelm Volkmann, *Anleitung zu den wichtigsten physikalischen Schulversuchen*, Rudolf Mückenberger, Berlin, 1912; Adolf F. Weinhold, *Physikalische Demonstrationen Anleitung zum Experimentieren*, Von Quandt & Händel, Leipzig, 1881; Eilhard Wiedemann, Hermann Ebert, *Physikalisches Praktikum*, F. Vieweg und Sohn, Braunscheweig, 1904; Johannes Wiesent, *Physikalische Vorlesungsexperimente Anleitung zur Ausführung der wichtigsten Versuche im Physikunterricht an Hochschulen un höheren Lehranstalten*, Ferdinand, Stuttgart, 1927; Lewis Wright, *Light a Course of Experimental Optics*, Macmillan & Co., London – New York, 1892.

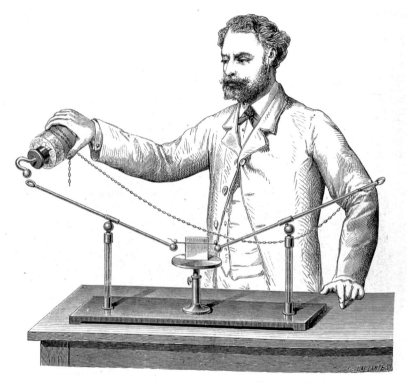

FIGURE 2.2   An engraving illustrating how to blow up a metallic wire with electricity.
AMÉDÉE GUILLEMIN, *LES PHÉNOMÈNES DE LA PHYSIQUE*, L. HACHETTE ET C.IE, PARIS, 1868, P. 374

behind a demonstration, but is fundamental from an experimental point of view.

Historical "brass and glass" instruments are fascinating and beautiful artefacts, as well as desirable collectibles, but certainly were not user-friendly. For example, even the apparently simple reproduction of a classical Newton's experiment such as the production of a solar spectrum with a prism can be problematic. If you do not want to obtain a blurred coloured spot of light but instead produce a neat and vivid spectrum band, you have to try several times: how to place the prism, to adjust the size and width of the slit, and so on.

Rarely will old treatises tell you such crucial aspects as how strong the currents in an electric apparatus have to be; or the distances among the various elements of an optical bench that are required in order to obtain the desired results. Sometimes adjustments can be done relatively easily and fast. But for more complicated experimental demonstrations, there can be many parameters to optimise and the preparation requires quite a lot of trial and error which can be really time-consuming.

FIGURE 2.3 The Author is about to explode a metallic wire.

In the past, high school and university physics professors who were teaching experimental physics were helped by the so called "*préparateurs*" (laboratory assistants). These experienced assistants knew the instruments and their workings extremely well. They prepared and tested the demonstrations until everything worked perfectly, so that professors found the apparatus ready to be presented in front of the classroom.

Finally, in order to successfully reproduce and film classical experiments, it is necessary to rediscover the practical know-how and the tacit-knowledge of these laboratory assistants. For these reasons, repeating and filming a demonstration with historical instruments can be compared to preparing a magic trick to be performed on stage. It will be successful and completely satisfying for the audience if they do not see the long preparation and the tedious training hours necessary to do it.

5   How to Select and Film Historical Experiments with Original Instruments

For various reasons not all historical instruments are suitable for being shown nowadays in a video. For example, various experiments (such as those involving high pressure gases) have to be avoided because the risk of accidents capable of seriously damaging the instruments is far too high. Furthermore, other thermometric or calorimetric experiments often require a long time (for reaching thermal equilibrium) and yet there is very little to show apart from the tedious movement of a mercury column in a thermometer. Videos illustrating them will be boring, and will not teach much about the working of instruments. For these and other reasons, we have to favour experiments and instruments which are quite spectacular. Furthermore, videos should not be too long: a few minutes are generally enough, and their aesthetic quality has to be professional. On the web it is possible to see thousands of videos with experiments made with modern or historical instruments, whose quality is pitiful: bad framing, awful light, blurred images, and non-existent scripts. Also, even if the content of some of these videos may be interesting and scientifically sound, their aesthetic quality is too bad to be acceptable, and therefore they are not very captivating. A professional video maker knows the rules and the conventions of cinematography and these often make a great difference. Here I give a short list of some of the most important criteria for selecting the instrument and the experiment to be to be filmed:

a) The necessary instruments have to be in good and working condition.
b) The risks of damaging the instruments during the demonstration have to be reasonably low.
c) The demonstrations have to be interesting from a scientific and historical point of view.
d) The demonstrations have to be spectacular enough to be filmed and presented in a few minutes.
e) Instruments should not have to be altered or modified in order to perform the demonstrations.

After having chosen the instruments, it is necessary to perform the desired experiment or demonstration and to prepare the set, in order to see that everything is ready for shooting. The demonstrator has to be sure that everything is in working order and that the various instruments are disposed on a table (or in the room) in such a way that they can easily be filmed. For our videos I and my team were lucky because we could always use different spaces (lecture room, collection main hall, laboratory, and so on) which were perfectly preserved like they were around 1900 (Fig. 2.4). So, we profited from natural and original settings (with old furniture, experimenting table, lamps, fittings, and so on) which are contemporary with the instruments to be filmed. We generally filmed with artificial light (apart from a few videos shot outdoors). The number of lamps for the set were reduced to a minimum, and with relatively simple equipment, our video maker was able to perfectly illuminate the set. We have to consider that today, video cameras, photo cameras, as well as Smartphones, not only are much cheaper, lighter and more user friendly

FIGURE 2.4   The author and Anna Giatti prepare Duboscq's lantern with the arc light and the dissolving view apparatus. Note the original late nineteenth-century furniture of the physics cabinet.

than the comparable equipment which would have been necessary a few decades ago, but they also produce images of high definition.

Generally, after having chosen the demonstration to be filmed and selected the necessary instruments, we wrote a simple plot for the video. After that scenario was written, the demonstration was first performed. It was filmed in its entirety once, or a few times, with a sequence shot. Later, following the indication of the performer, every single step, such as connecting the cables, lighting a lamp, closing a switch, rotating a crank, looking in an eyepiece, and so on, and every constructive and working detail of the instruments, were filmed, one after the other.

In order to reproduce as accurately and precisely as possible the different experiments presented in our videos, we wanted to avoid any disturbing anachronism. In certain cases, it was very easy (using old fashioned matches instead of a modern lighter to ignite a flame, chemical substances were stored in old chemical bottles and not in modern plastic containers, for instance). In others, avoiding anachronism was more complicated, and more work was required. For example, when we needed a strong light source (for optical experiments, projections, spectroscopy) we used an original arc lamp with a regulator which required a powerful direct current (DC) source. High voltage necessary for

FIGURE 2.5 The author and Roland Wittje experimenting with an electric arc during the filming of the "speaking arc" demonstration. The arc was produced thanks to eight car batteries.

electrostatic demonstration, or for illuminating Geissler tubes, was produced with a Wimshurst electrostatic generator or with an old induction coil. In our filming, old Bunsen burners were used together with spectroscopes, and so on. We simply imagined ourselves to be in a 1900 laboratory and to use only instruments and tools which could be commonly found at that time.

In only a few cases was it impossible to avoid using modern technology. In filming the demonstrations of speaking arc and singing arc, we needed a strong source of DC (Fig. 2.5). At the beginning of the twentieth century, DC was commonly supplied by the network, or every laboratory of a certain importance had large electrochemical batteries which were generally charged by a dynamo machine. Unfortunately, we did not have a high DC source at our disposal. We tried to use alternate current (AC) and rectify it, but our commercial rectifier did not completely eliminate the 50 Hz pulsation, which produced a very loud attendant noise in our electro-acoustic system. So, we decided to buy eight modern car batteries, which proved to be the ideal solution (but do not appear in the video). In the demonstration of the Koenig acoustic siren, we needed a strong jet of air. But the two old blowers with bellows, in the historical collection, were far too fragile to be used. A modern compressor gave us a lot of pressure, but its air flow rate was too low. Finally, we solved this problem with a few scuba tanks that we rented in a local sports store. With them, we could modulate the air flow quite easily without having the disturbing noise of a compressor.

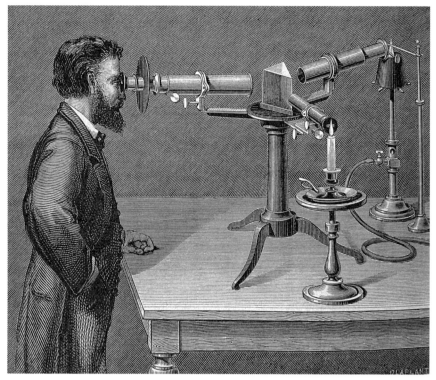

FIGURE 2.6   A nineteenth-century engraving illustrating how to use a spectroscope. Note that the size of the instrument is far too big compared to that of the person.
AMÉDÉE GUILLEMIN, *LES PHÉNOMÈNES DE LA PHYSIQUE*, L. HACHETTE ET C.IE, PARIS, 1868, P. 660

Several problems arose when we had to film the images which could be seen in the eyepiece of a spectroscope, a polarimeter, or a reading telescope. Because of our very limited budget we were unable to acquire high quality digital microscope cameras, while the images obtained with cheap ones were not satisfying at all. Finally, we noted that the best results were obtained by using the camera of a good Smartphone, whose objective lens was set against the eyepiece of the apparatus to be filmed (Figs. 2.6 and 2.7) Also, even if the colours of the images of the spectra did not correspond faithfully to the visual perception of the human eye, they were good enough for our didactic videos.

With our well-trained team it was possible to film up to six or seven demonstrations in five working days. That was possible because Anna Giatti and I had perfect knowledge of the instruments and how to manipulate them. On the other hand, after long experience with us, Antonio Chiavacci knew well what we wanted to show and he needed very few indications from us.

FIGURE 2.7   The Author using a nineteenth-century spectroscope. The gas flame illuminates its divided glass scale, whose image is superimposed on the image of the spectrum. The other lamp is the light source for producing the spectrum.

6   Postproduction

After the shooting in the physics cabinet, there was the post production. The video maker edited the images and sent a first version of the clip to me and Anna Giatti. Generally, a few corrections and modifications were necessary: certain frames were too long or needed to be cut, while certain details, which appeared to be necessary for a thorough comprehension of the instruments, had to be added using the filmed material. Generally, the second or third edit version was considered good, next it was possible to add the captions. In fact, most of our videos are silent, or better, only reproduce the sounds and noises of the demonstrations. The operators do not speak, nor is there a voice commentary added during the post production.[7] This was a choice dictated by two criteria: simplicity and economy. Considering that we wanted the maximum diffusion for our videos, it would have been necessary to hire a professional speaker with English as the mother tongue because a spoken Italian

---

7   That is not true for the videos made in the high school in Bergamo. For them our sponsors required to have an Italian commentary read by a speaker.

commentary would have been understood only by a small percentage of the viewers. But a speaker would have been too expensive for our budget and would have slowed down the post production. Finally, we decided to insert under the images a series of captions in English, while the same captions (in Italian) were added to the page under the video themselves in YouTube. This solution proved to be quite satisfactory. I do consider that a few captions, or a very simple comment, are enough for explaining what is going on in the video. In the commentaries, we avoided giving interpretations, historical remarks or critical comments on the experiments. We did not want to teach physics nor history of science and technology; we just show and explain how some classical instruments work and what they are used for. In such a way, the videos remain totally "open" in the sense that they can be used in several different occasions and by different people or different institutions. Two videos were made in collaboration with colleagues who needed to repeat some particular experiments: we filmed with Roland Wittje various experiments with the singing and speaking electric arc, and with Eugenio Bertozzi a working Wilson cloud chamber of the 1950s.

All our videos, the ones made in the Fondazione Scienza e Tecnica in Florence as well as the ones of the Liceo Paolo Sarpi of Bergamo, were uploaded to the YouTube platform and are open access. In fact, our videos, which are mostly sponsored by public money or thanks to non-profit institutions, can be used and distributed freely.[8]

7 Uses and Users

Our short videos can be used by different people and in different contexts. An immediate and local use is to provide more information for the visitors of the collections which preserved the filmed instruments. In the Fondazione Scienza e Tecnica of Florence, a label with a QR code has been placed near the instruments appearing in the videos. The visitor can scan the code with a smart phone or a tablet and is immediately directed to the YouTube page showing the working of the instrument. In several cases, in the same video it is possible to see several instruments used conjointly for a single demonstration (for example an electrostatic machine, a battery of Leyden jars and a "thunder house").

---

8 Where the videos are shown in museums, exhibitions, educational institutions, etc., we only ask that they are used without further modification.

Several museums and collections around the world preserve scientific instruments which are identical, or very similar, to the ones presented in our videos and therefore they can use our videos in their institution for explaining their own instruments. For example, Teylers Museum in Haarlem as well as the Electropolis Museum in Mulhouse, France, just to mention two of them, asked to use our videos in their institutions. Furthermore, historians of science and historians of instruments use our videos in their courses for illustrating some classical experiments or for showing some particular apparatus. Finally, physics teachers and professors can also find them useful. Many modern instruments, while extremely performance- and user-friendly, are often sealed and impenetrable black boxes, typically connected to a computer. From a didactic point of view, showing the Faraday effect with late nineteenth-century instruments or producing some optical spectra by using classical apparatus is often more effective and illuminating than working with the equivalent modern equipment.

Apart from these professional and institutional users, the videos are regularly seen by students, amateurs, collectors, restorers, and others. Considering that we are filming very peculiar historical artefacts which can interest quite a small number of people, it is impressive that, up to August 2021, over 711,000 views were recorded in the YouTube channel of the Fondazione Scienza e Tecnica. It would require a few decades to attain the same number of visitors in the collections where the videos were shot! It also has to be pointed out that certain videos, like the ones showing some of the Hertz experiments with electromagnetic waves and with very early wireless demonstration apparatus, or the Faraday effect, are particularly popular and were seen several tens of thousands times (Fig. 2.8). Certainly, it would be interesting to analyse the typology of our viewers, the duration of their virtual stays, their geographic distribution, the reasons why certain videos are more popular than others, etc. These parameters could be studied and presented in a further article. Now the YouTube channel of the Fondazione propagates our videos without the need for our intervention. It just requires a little attention in order to avoid the posting of improper comments, to answer to the questions of the followers, and to profit from the possible useful relations and connections, which can be developed thanks to wide public access to the videos.

FIGURE 2.8 A frame of the video dedicated to the Faraday effect. This video was seen about 50,000 times between December 2012 and October 2020.

8 Conclusions

The success of our videos and the many positive comments left on social media are certainly a reward for our work and encourage us to carry on. We really believe that our work needs to continue: more interesting instruments remain to be filmed. Unfortunately, for many reasons, the situation in the Florentine collection is now much less favourable and we do not know where and how we will be able to continue to film. However, I just hope that our videos (and this article) can be useful, not only for illustrating and better explaining the working of some instruments of classical physics, but also for stimulating other people and other institutions to produce their own videos.

CHAPTER 3

Making It about the Objects: A Reboot of a History of Science Course

Janet Laidla | ORCID: 0000-0003-4055-8845

1   Introduction

This chapter introduces a collaboration project between the Institute of History and Archaeology and the University of Tartu Museum at the University of Tartu (Estonia).[1] Through this project, a new elective course in Estonian history of science was created that not only introduces history of science and scientific instruments to (humanities) students but also gives basic knowledge in object handling, photography, and graphic design.

History of science is part of several introductory courses at the University of Tartu, such as history of medicine, pharmacy, psychology, geology, physics, mathematics and others.[2] Sometimes as part of these courses the lecturers bring the students to see the collections at the university museum, but this is not usually a very hands-on experience. Some of the departments have their own collections, such as the Department of Physics, and they do use their objects for demonstrations.

There used to be a *General History of European Science* course at the Department of General History, but this has not been on offer for five years. So there came about the idea of creating a new object-centred course. To be more precise, the course should be called *History of Estonian Knowledge* as the humanities were included as well.[3] However, scientific instruments and teaching aids dominated in the final choice of items.

---

1   This chapter is based on a presentation "Making It about the Objects: A Reboot of a History of Science Course", presented at the *XXXIX Symposium of the Scientific Instrument Commission*, London, UK, 14–18 September, 2020. The collections involved in the preparation of this course were introduced via collections tours during the *XXXIII Symposium of the Scientific Instrument Commission*, Tartu, Estonia, 25–29 August 2014.

2   Search in the Study Information System (SIS) of the University of Tartu: https://ois2.ut.ee/#/dashboard (accessed 19 Jan. 2021); the guest version might not offer the same results as the teacher version.

3   In Estonian *Eesti teaduse ajalugu*. The word *teadus* includes exact, natural sciences and humanities similarly to German *Wissenschaft*.

© JANET LAIDLA, 2022 | DOI:10.1163/9789004499676_005

The activities of the course were driven by the following four main goals:

1. To use the collections of the University of Tartu Museum in university teaching, as these are not used as extensively as they could be;
2. To introduce humanities students to three-dimensional sources, as they mostly read and produce texts in other courses;[4]
3. To give students basic object handling skills they can use if they decide to have a museum related career;[5]
4. To give the students basic photography and graphic design skills that will be useful for their future career both inside and outside of academia.

Some authors have remarked that museums in general have targeted their educational activities mostly towards primary and secondary school children.[6] At the same time many university museums are working more closely with university students. The University of Tartu Museum has thus far been in this respect closer to a general museum than a university museum, as it has in the past endeavoured to build up a strong educational package offered mainly to primary and secondary school pupils. University level students have been included through numerous work placements and introductory tours. The goal is, however, to engage even more students through special courses.

Both topics relevant to this chapter – the object-based learning in higher education and the collaboration between the universities and museums – have over past decades been the subject of a growing number of publications on both general topics and specific case studies.

This chapter will introduce the participants, the parts of the course, the objects used in the course, the students' feedback, and will conclude with general remarks and analysis. As the course has been taught only once, the source material is not extensive and consists of the lecturer's personal experiences

---

4 This also applies to research, but is slowly changing, see for example: Chris Gosden and Yvonne Marshall, "The Cultural Biography of Objects", *World Archaeology* 31, 2 (1999), pp. 169–178: p. 169.
5 As Estonia is only building up museum studies curricula at the University of Tartu and the Estonian Academy of Arts in Tallinn, many new employees have never handled a museum object. The University of Tartu Museum has tried in-house training, and I am now expanding similar exercises to my courses.
6 Catherine Speight, Anne Boddington and Jos Boys, "Introduction", in A. Boddington, J. Boys and C. Speight (eds.), *Museums and Higher Education Working Together. Challenges and Opportunities*, Routledge, London and New York, 2016, pp. 3–23: p. 3. There is a similar situation in literature on museum education; see Helen J. Chatterjee, Leonie Hannan and Linda Thomson, "An Introduction to Object-Based Learning and Multisensory Engagement", in H.J. Chatterjee and L. Hannan (eds.), *Engaging the Senses. Object-Based Learning in Higher Education*, Routledge, London and New York, 2016, pp. 1–18: p. 1.

and remarks during the course, the voluntary student feedback form and the student posters.

2 The Students

Eleven students opted for the elective course. Of these, three were graduate students and eight undergraduate students. Most of the students were from the Institute of History and Archaeology, one from the English and one from the Psychology departments. The course was open to all the students; however, in the Study Information System (SIS) of the University of Tartu it appeared in the timetable of the Institute of History and Archaeology. The students are able to search the SIS for interesting elective courses from other faculties and departments. The number of students from outside the department that offers the course varies greatly depending on the specific course.

3 The Objects and the Activities

One of our first difficulties was to find suitable museum items in cooperation with all curators at the University of Tartu Museum. The students were both at undergraduate and graduate levels and had no background in the history of science. We did not know how proficient they would be in using nineteenth-century sources in German and Russian languages. We also needed to know there was potential for finding further information that was accessible to students. Thus, we needed just enough information about the object for the student to get started, such as the time period, name of the user or maker, but at the same time to choose objects that had very short or non-existent descriptions in the museum database so that the student can potentially make an original contribution.[7] The final selection included scientific instruments, teaching aids and some so-called everyday objects in our collections, such as the gas light and the office phone. Some instruments worked better than others did and that depended on how much new information the students were able to find and not on the nature of the object.

The course began with two general lectures on the nature of history of science in general (how it has been written, newer developments in history of

---

7 However, as this was the first time the course was offered in this way, some items in the final choice had pre-existing descriptions. In those cases, the student could not make a huge original contribution; rather the student specified and elaborated on the existing description.

MAKING IT ABOUT THE OBJECTS

science and a very general chronological overview of developments in science). These were followed by three reading seminars on the Estonian history of knowledge. The students had to choose two subject fields and periods and could decide between several reading options. The seminars were meant to give a short general overview of the most important developments in the history of Estonian knowledge between the early modern period and the 1990s.

After this introduction, it was time to move the course work to the museum. First, we held a general object handling session, covering basic knowledge on how to touch, move, describe and store an object. During the first session, a short introduction to all of the items chosen for the course was provided (name, date and function, if known) and the students could think about their choice of an object for further research. During the second session, the students had to finalise their choices and the remaining items were taken back to the storage. The students were working individually and most of them worked with one object of their choosing. They inspected their objects, made notes and took pictures with their phones. They were able to choose the topic they wanted to concentrate on (object, maker, use, wider background). More often this was actually later determined by the available information.

In a homework reading assignment, the students were provided with the official guidelines for photographing 3D items prepared by the Conservation and Digitisation Centre Kanut. During the photography session the basics on lighting (lights, using the light-box), background, shutter speed, ISO and aperture were repeated and the students were able to make high(er) quality photographs from the different angles they would use in their posters. As some of the items did not have an image in the database, the students' images could be added to the database at a later date.

During another seminar session, the students shared their findings and received questions and feedback from other students. These responses would direct their further research and the contents of their poster. As they presented their findings in seminars, there was a general sense of excitement and pride that they had actually uncovered new information about the objects for the museum.

There were two voluntary workshops in the computer laboratory for those who were apprehensive about their graphic design skills. Basics of poster design with some examples and relevant software were introduced.[8] The students then began sending in the poster drafts to the instructor and these were

---

8 History students and historians in general tend not to use the poster format for presenting their research in Estonia. However, it is a good opportunity for students if they want to participate in larger international conferences in the future.

edited (sometimes several times) before printing. The instructor made suggestions on both layout and content and pointed out typos. Suitable exhibition space in the stairway of the University of Tartu Museum was found and a spontaneous exhibition opening was arranged with some food and drink after the course had officially ended in June. Most of the students and the curators participated in this opening, providing the students with very positive feedback from the museum staff.

With their research, the students were able to enhance the existing object information by adding makers, specifying names of the objects and how the instruments were used, and by adding historical background information to the museum database.[9]

Altogether, I chose twelve objects for the course from around thirty objects proposed by the curators. These twelve gave a variation of medical, anthropological and psychological instruments; teaching aids for physics, chemistry, geology and art; a gas light and an office phone. The two objects the students did not choose were two-dimensional: the architectural drawings and honorary dedication for a professor of medicine, Artur Linkberg (1899–1970).

One of the students worked with a number of related objects. These were the models of molecules created by Estonian doctor Raik-Hiio Mikelsaar (b. 1939) (University of Tartu Museum, henceforth ÜAM, 1377:1–7 AjFK), the molecule building sets (ÜAM 1682:1–2) and the tools for assembling the models (ÜAM 1682:3–6). The models of molecules were a separate earlier acquisition; later the museum was also able to acquire two sets of building pieces and the set of tools for assembling the models. The student concentrated on these later objects. As Dr Mikelsaar is still living, one option would have been interviewing him, but the student began at the archive of the museum to study the historical publications introducing Mikelsaar's models to gain more insight,[10] and felt that no further information was required for the poster.

Another teaching aid, the galvanometer (ÜAM 320:1 Aj), was researched by a student who concentrated on its history and working principles. The object had also been described in quite some detail by our curator of the physics collection. The student was able to build on that description and contribute good quality images to the museum database.

---

9   Estonia has a national collections database, Information System of Estonian Museums, which also has a Museums Public Portal: www.muis.ee/en_GB (accessed 28 Oct. 2020). Most of the Estonian museums have joined this database, which is a collections management tool that can be used for loans, exhibitions, inventories, conservation notes, etcetera.

10   Raik-Hiio Mikelsaar, *New precision space-filling atomic-molecular models*, USSR Academy of Sciences. Pushchino, 1985.

MAKING IT ABOUT THE OBJECTS 55

FIGURE 3.1   Craniometer or, according to the Baltic-German anatomist Eber Landau, a cephalograph
UNIVERSITY OF TARTU MUSEUM 1110:5 AJM 113:5; PHOTO BY MART KÜNG

The student who chose the adenoid curette (ÜAM 1576:21 AjM 198:21) concentrated on Ludvig Puusepp (1875–1942), the professor of neurology, as the curette came from his clinic. The second part of the poster was on the local Tartu instrument maker Aleksander Paul Keiss who had made the curette.

A Graetzin-Licht gas light (ÜAM 12:24 AjFK 8:24) was one of the older acquisitions and had been in the collection since 1984. A number of gas lights from different buildings of the university have ended up in the museum when the buildings were emptied for renovation. This particular lamp came from the department of inorganic chemistry. The student was able to re-date the lamp, establishing it as a fixture from the early twentieth century, instead of from the nineteenth century, and to provide background information on the lamp company.

There was an acquisition from the department of psychology from 1992 that had thus far escaped the curators' attention and one of these instruments was suggested by the curator for further research as these were to be photographed and identified.[11] The student with the background in psychology was able to identify the instrument as a tachistoscope (ÜAM 700:11Aj). Information about its use and maker were included to the database.

A controversial early anthropological instrument had been thus far identified as a craniometer (ÜAM 1110:5 AjM 113:5, see Fig. 3.1), but the student noted

---

11   The general identification for these objects in the records was "Instruments of psychology".

FIGURE 3.2 Louis Ombrédanne's anaesthesia mask by Mathieu & Gentile Collin and H. Windler
UNIVERSITY OF TARTU MUSEUM 991:1A; PHOTO BY GERT KLAASEN

that its maker, the Baltic-German anatomist Eber Landau (1878–1959), referred to it as a cephalograph. The database record was improved with the use, maker, and the new term and references to the articles where Landau talks about the instrument and his research. New higher quality photos of the item were also uploaded to the database. The student also identified a collection of photographs at the Estonian National Museum that is connected to the instruments held at the university museum.

The student investigating the Louis Ombrédanne's anaesthesia mask (ÜAM 991:1a, see Fig. 3.2) was able to improve the database record with information that the mask was made in two factories. The part that goes over the mouth and nose was made by Mathieu & Gentile Collin, Paris, and the remainder at H. Windler in Berlin. The student also provided information about the French surgeon Louis Ombrédanne (1871–1956). This student noted that most of his results were not included on the poster because of the limited space.

Another student who was surprised to find out that all the new information could not fit in the poster, was the one who chose an office phone (ÜAM 1660:1Aj)! It belonged to pharmacologist and Vice Rector for Research

MAKING IT ABOUT THE OBJECTS

Johannes Tammeorg, working at the University of Tartu between 1940 and 1986. It turned out that Tammeorg was involved in so many organisations and was so well connected that he definitely needed a sophisticated modern telephone. The student added information on the factory and about the functions of the phone.

Professor of Chemistry Carl Schmidt (1822–1894) left a cupboard with drawers filled with different rocks and minerals. Curators chose a collection of nails (ÜAM 945:1080 AjM 70:1080) for the students to investigate. The student who took on the task was happy to report that the puzzling collection of nails was used for the Mohs hardness test and therefore it was logical that it came with the mineral collection.

The only item that clearly represented the humanities was the model of the Pyramid of Caius Cestius (University of Tartu Art Museum, KMM TK 116). The student delved into the wider context of the item – the popularity of making models out of cork coinciding with interest in Egyptian antiquities. The pyramid also characterises the popularity of Egypt in Rome during the building of the original (between 18–12 BC). Made in the beginning of the nineteenth century by Friedrich Heubel in Schwarzburg, Germany, and bought by the Professor of Classics and the founder of the University Art Museum Johann Karl Simon Morgenstern (1770–1852), it was probably used as teaching aid.

In summary, the students learned about the history of Estonian and European knowledge through their background investigation into the makers, users and wider context of the objects. They also learned to use new types of sources, not only the objects themselves, but the scientific publications and instrument catalogues. They learned the basics of object handling and about the challenges connected with presenting their findings in an aesthetically appealing way on a poster.

4   Feedback from the Students

After the course, a voluntary feedback in Google-form was used. The questionnaire had ten questions, of which four required the student to score different parts of the course on a scale from 1 to 5 (lectures, seminars, hands-on and graphic design sessions).[12] The other six questions required the students to comment: did the course meet their expectations, how did the students feel

---

12   1 = Not at all interesting and/or useful; 5 = very interesting and/or useful. In a feedback form for another pilot course the evaluations on "useful" and "interesting" were given separately and it did reveal an existing but slight difference, thus this should also be done for the feedback form next time.

about creating a poster, what did the students think of courses that included both theory and hands-on activities, what would they do differently, what other skills would they like to acquire, and any additional comments.

As this was a voluntary questionnaire (the students usually have to fill a mandatory central feedback form, thus it is not common to make these extra questionnaires mandatory) only six students completed it. However, in general, those participants, who feel strongly either way, are more likely to respond to the questionnaires.

The object-handling workshop received the maximum score from all participants. The lectures, seminars and graphic design workshops received scores from 3 to 5. Two students gave maximum scores for all the activities. There seems not to be a pattern in the less-highly rated activities, so it can be suggested that students' previous knowledge and personal interests or preferences (learning styles) may have influenced these scores.

Oral feedback was also collected during the course and during the opening of the exhibition. Overall, the students were pleasantly surprised with the practical side of the course, but a few students mentioned that the general theoretical background could have been more substantial – thus it is important to find a balance between the theoretical background information and the practical object-driven activities.

The graphic design workshop was voluntary and the students did not consider it equally useful either because it was too primitive (if they had previous experience) or too limited (if they had no previous experience at all). The students' feedback suggests that the graphic design workshop should be kept voluntary and specific learning outcomes should be provided so that the students know what to expect.

The students valued highly the opportunity to present their work both orally and visually as it was different from their usual coursework. Two students mentioned this aspect specifically in the feedback questionnaire. They also valued the fact that the presentation of their results was public and their information was used by the curators; this was mentioned by one student.

However, they also noticed that this was the first time this course was offered and that the way time was divided between different tasks could be improved upon. Firstly, again the balance between theoretical courses[13] and

---

13   The overall feedback shares some similarities to the results of a study conducted at University College London between 2012 and 2014. Notable is the students' need for a better connection between wider theoretical context and the object-handling session. See: Arabella Sharp, Linda Thomson, Helen J. Chatterjee and Leonie Hannan, "The Value of Object-Based Learning within and between Higher Education Disciplines", in Chatterjee and Hannan (eds.), *Op. cit.* (n. 6), pp. 87–116: especially pp. 107, 111–112.

MAKING IT ABOUT THE OBJECTS

practical workshops, and, secondly, individual photography sessions in a very limited space left most of the students just sitting around and chatting when they could have done something else.

This (probably a graduate student's) comment perhaps summarises the overall feedback: "I felt that this course was read for the first time in this way: good ideas, but the implementation was a bit rocky. At the same time, I'd say it was my favourite course so far. Every semester should include this kind of hands-on course."

5   General Observations

When preparing for the course, I did not have a specific general theoretical (that is pedagogical) background in mind. My main source of inspiration had been object-handling sessions provided by numerous university museums. As the students enrolled in the Tartu course worked with the objects for longer than one session, the object-handling session grew into object-based research, which is introduced and described in several papers in *Winterthur Portfolio*.[14]

In addition to using objects, my goal was to enhance the cooperation between the university teaching staff and the museum. When museums and universities collaborate they can provide complementary resources that may enhance the learning experience as well as learning with objects can enhance deep learning.[15] Museums can also be the centres for object-based research and thus provide their expertise in working with this type of source. Working directly with objects is one way of making the museum experience interactive and that, in turn, provides a sense of involvement with the collections and also a feeling of enfranchisement[16] or ownership.[17] During the presentations in this

---

14   See for example Jules David Prown, "Mind in Matter: An Introduction to Material Culture Theory and Method", *Winterthur Portfolio* 17, 1 (1982), pp. 1–19; Edward McClung Fleming, "Artefact Study: A Proposed Model", *Winterthur Portfolio* 9 (1974), pp. 153–173.

15   C. Speight, A. Boddington and J. Boys, "Introduction", *cit.* (n. 6), p. 4.

16   Jerome de Groot, *Consuming History. Historians and Heritage in Contemporary Popular Culture*, Routledge, London and New York, 2009, p. 246. This can also be connected with the history of touch in the museum. Fiona Candlin suggests that touch in museums has been exclusive and reserved for the privileged visitors: Fiona Candlin, "Museums, Modernity and the Class Politics of Touching Objects", in Helen J. Chatterjee (ed.), *Touch in Museums. Policy and Practice in Object Handling*, Berg, Oxford and New York, 2008, pp. 9–20.

17   See for example Kirsten Hardie, "Engaging Learners through Engaging Designs that Enrich and Energise Learning and Teaching", in H.J. Chatterjee and L. Hannan (eds.), *Op. cit.* (n. 6), pp. 21–41: p. 23.

rebooted course, many of the students spoke of feeling a stronger connection with the object they had to research. This feeling of ownership may increase their motivation and also the level of deep learning.

Learning that involves objects (object-based, object-centred or object-inspired learning) has been connected to acquiring both thematic and inter-disciplinary information, and also gaining practical and transferable skills, such as observation and inquiry.[18] Helen J. Chatterjee, Leonie Hannan and Linda Thomson have summarised many of the existing learning theories rel-evant for the object-based learning experiences and there are several inter-esting points. For example, Shawn Rowe's scenarios of knowledge which are constructed in group or social situations are relevant to the activities of this course.[19] Although each student had individual items, the activities in general were more social and built around one table or one camera, rather than a tra-ditional lecture session. As students chose the different aspects they would concentrate on, and had to introduce their findings in a seminar, they were also able to receive feedback from other students. The students read each oth-er's posters with deep interest and became aware of other aspects which were relevant in studying scientific collections, even if they had decided not to con-centrate on these specific aspects for their projects. In some cases (and this can be improved) the knowledge was created in discussions between the student and the curator, and between the students enhancing the social aspects of the course.

Although in several other object-based courses the students are building, copying or using replicas to gain more insight,[20] the students' interaction with the objects on this course can also be considered active and stimulating.[21] They were not able to create a copy but they were able to (carefully) inspect and handle the objects. Perhaps the most stimulating aspect was the unusual environment and sources that did not yield information as freely as many texts do. Serial numbers, maker's logos, geographical locations and other markings were sought out and their meaning contemplated by the students. In some cases, identifying the object required very careful comparison with images of other similar instruments.

---

18   Chatterjee, Hannan and Thomson, "An Introduction ...", *cit.* (n. 6), p. 1.

19   *Ibid.*, pp. 3–4 referencing Shawn Rowe's 2002 publication "The role of objects in active distributed meaning-making" and L.S. Vygotsky's 1978 book *Mind in Society: Development of Higher Psychological Processes*. See also Nina Simon, *The Participatory Museum*, Museum 2.0, Santa Cruz (CA), 2010, pp. 127–181.

20   As in many case studies in this volume.

21   Chatterjee, Hannan and Thomson, "An Introduction ...", *cit.* (n. 6), pp. 2–3 on the theories of George E. Hein's 1998 book *Learning in the Museum*.

MAKING IT ABOUT THE OBJECTS   61

In addition to the aspect of object-handling, the course experience was definitely also inquiry-based learning, as each object posed some kind of conundrum for the student and the solutions were not readily offered – or even known! – by the lecturer. Thus I can attest that during this course, the lecturer became the facilitator (more than usual, I feel),[22] as the students were able to carry out research individually. Inquiry-based learning has been defined in various ways, but the common notion is that it should include some kind of problem (in this case, the object) that the student has to solve.[23] Based on previous research, Anindito Aditomo, Peter Goodyear, Ana-Maria Bliuc and Robert A. Ellis introduce three forms of inquiry-based learning: problem-based (business education), project-based (design) and case-based (medicine).[24] I would argue that object-based learning would be an additional form,[25] which has been practiced by researchers writing object biographies:[26] it would start with an object, and the students would study different aspects of the object using the object itself and additional background materials.[27] Similarly to problem-based and case-based teaching, the pedagogical emphasis is on process and the purpose is to acquire new knowledge. This specific course also added the project-based learning element on top of object-based learning as the students needed to design a poster. The inquiry-based assignments could be developed more thoughtfully to improve students' general research skills and promote object-based research.

One of the differences between the case study in this chapter and some others in this volume is that in this course there were mostly humanities students studying instruments or teaching aids of science and medicine. Humanities students often engage with archaeological, historical and ethnographic collections. Scientific instruments are commonly used in courses for engineering or

---

22   See *Ibid.*, pp. 5–6, referencing James Oguneye's 2007 book *Guide to Teaching 14–19*.
23   Anindito Aditomo, Peter Goodyear, Ana-Maria Bliuc and Robert A. Ellis, "Inquiry-based Learning in Higher Education: Principal Forms, Educational Objectives, and Disciplinary Variations", *Studies in Higher Education* 38, 9 (2013), pp. 1239–1258: pp. 1240–1241.
24   *Ibid.*, pp. 1241–1242, and Table 1.
25   Almost similar to problem-based as the students were facing a real-life problem of a museum curator encountering an item in collections they have little information on.
26   See for example: Gosden, Marshall, *Op. cit* (n. 4) for an example in humanities, and David Pantalony, "Biography of an Artifact: The Theratron and Canada's Atomic Age", *Scientia Canadensis: Canadian Journal of the History of Science, Technology and Medicine / Scientia Canadensis: revue canadienne d'histoire des sciences, des techniques et de la medicine* 34, 1 (2011), pp. 51–63 for the scientific instruments. These are good examples for the students; however, the short course might not give enough theoretical background for such a substantial study.
27   See the Winterthur model in Prown, *Op. cit.* (n. 14), pp. 7–10.

science students who are learning science with the aid of objects or replicas. As part of their reading seminars and research, the students gained insight into other fields of knowledge and other ways of creating knowledge that might help them to gain a better understanding of other subject fields.

As object-based activities are more time-consuming to create in comparison to traditional lectures and seminars, time is a relevant factor. From the key challenges identified by Chatterjee, Hannan and Thomson,[28] the lack of time here falls under the economic rather than the cultural challenge. However, cultural challenges exist as well. The problem of location (that is, collections not on site) is not as challenging for us, as the University of Tartu is situated in a small city with numerous museums and excellent collections. Another economic issue is also the relatively small (but existing) material costs connected to physically creating something visual, in this case printing the poster exhibition and holding a celebratory opening. It should be decided early on who will cover these costs to avoid any tension later.

6 Areas of Improvement

The *History of Estonian Knowledge* course is on offer again in spring semester of 2021.[29] The first set of objects was chosen in the autumn semester by a student intern as part of her internship. The curators will add their suggestions to this list. The lecturer is now able to provide a more specific call for suggestions: objects with little description and no suitable image. However, the items on the list should provide clues about the objects on them (the maker or name) or in the museum's acquisition records.

More emphasis will be given to the general theoretical overview on the history of Estonian knowledge with two additional seminar times. The photography sessions will be arranged in smaller groups of a maximum of two to three students at the time to address the feedback from the first pilot year of the course. After the students have chosen the item they will work with, the object handling session will include a worksheet inspired by the Winterthur model. This is based on the fields of the Museum Information System to systemise

---

28 "Logistical, economic and cultural" in Chatterjee, Hannan and Thomson, "An Introduction ...", *cit.* (n. 6), pp. 9–11.

29 This course is part of an action research project that is part of my work for the grant in scholarship of teaching and learning. On action research see for an overview: Marie Paz E. Morales, "Participatory Action Research (PAR) cum Action Research (AR) in Teacher Professional Development: A Literature Review", *International Journal of Research in Education and Science* (*IJRES*) 2, 1 (2016), pp. 156–165.

MAKING IT ABOUT THE OBJECTS

the way the students will observe the object in the hope that it will improve their observational skills.[30] The lecturer will analyse the potential research skills (both general[31] and object-based research) and tweak the assignments in a way that the course work would also systematically improve students' research skills.

Instead of two sessions of graphic design, the computer laboratory could be used to supervise the students in filling out the museum information database with their own findings. Students would be issued temporary user accounts to do this task, and curators' time would be spared. A more flexible graphic design workshop will be considered that could cater for the needs of both beginners and more advanced designers.

To gain greater variety of background knowledge among incoming participants, new ways of attracting students from different departments will be considered. However, the maximum number of participants should not exceed sixteen in order to allow each student adequate time to give a presentation during seminars and ensure enough room for the object handling sessions.

7   Concluding Remarks

The reboot of the Estonian history of knowledge course can be considered as successful, but it can be improved. Creating a well-balanced theoretical course that also introduces several different practical skills is definitely more time-consuming and complicated than doing a theoretical lecture course, but it is worth the effort. Both the students and the museum appreciated the result. This first run of the object-led history of knowledge course provided the lecturer and the museum confidence that from the museum's side university students can be meaningfully integrated into object-based research and from the university's side that these object-based activities are both stimulating and useful for students and help to achieve relevant learning goals.

Student motivation increased through participating with real objects, such as scientific instruments, teaching aids and other kinds of academic heritage,

---

30   With references to additional comments to the model made from the scientific instrument point of view, see Katharine Anderson, Mélanie Frappier, Elizabeth Neswald and Henry Trim, "Reading Instruments: Objects, Texts and Museums", *Science and Education* 22 (2013), pp. 1167–1189.

31   A helpful model is provided in: Angela Brew, "Understanding the Scope of Undergraduate Research: A Framework for Curricular and Pedagogical Decision-making", *Higher Education* 66 (2013), pp. 603–618, and Fig. 3.

and through creating posters for the real exhibition at the museum. Real gains were made in the museum database; however small those additions might be.

The learning experience improved by introducing object-inspired problem-based learning assignments that are similar to a curator's real-life work. In addition to providing object-handling and graphic-design skills that are not common to history courses, the assignments also have a potential of improving students' research-skills.

New and enhanced learning opportunities can be created when museums (libraries, archives) and the university lecturers co-operate. This will require extra effort and resources in the beginning that should lessen somewhat when cooperation becomes an established tradition.

Acknowledgments

I would like to thank Elizabeth Cavicchi and Peter Heering for their guidance and numerous suggestions that improved the quality of the paper immensely. I would also like to thank the participants of the *XXXIX Scientific Instrument Symposium*, London, 14–18 September 2020, for their questions and comments that improved this chapter. Mari Karm from the University of Tartu also provided useful notes and links to relevant literature. This chapter would not exist without the wonderful students who took the course in Spring 2019 and the curators at the University of Tartu Museum.

CHAPTER 4

Using Original Instruments from a Museum Collection in Demonstrations

Jan Waling Huisman | ORCID: 0000-0002-1332-3218

1 Introduction

In this chapter, I give an overview of the educational programmes that make use of scientific instruments (originals or replicas and reconstructions), performed in the Museum of the University Groningen (UMG), the Netherlands.[1]

As is the nature with educational programmes in a museum, these vary from year to year, and sometimes are specially made and performed for single occasions. In addition, because of some specialised instrument (collections) within our collections, there are scientists wishing to perform experiments with original instruments who come to us.

Finally, yet importantly, we use instruments to brighten up congresses and presentations. Our "Wagentje van Stratingh" (the oldest known electric car still in existence, with provenance) is in especially high demand for anything that has to do with green energy or sustainability; so much so, that we have had a replica of it made, which the University's Nobel laureate Ben Feringa often uses during presentations around the country.

In this chapter, I also give an overview of the history of scientific instruments of the University of Groningen, followed by some cases explaining their educational use and, finally, some notes on what we can learn from such a use.

2 Some Historic Instruments of the Collections

The year 1698 was an important one for scientific instruments at the University of Groningen. Johann Bernoulli (1667–1748) then professor of mathematics (1695–1705), succeeded in buying a state-of-the-art and very expensive

---

1 Some of these programmes have been discussed in: *SICU2 – An International Workshop on Historic Scientific Instrument Collections in the University*, Oxford (MS), USA, 21–24 June 2007; *XXXVIII Symposium of the Scientific Instrument Commission*, Havana, Cuba, 23–27 September 2019.

© JAN WALING HUISMAN, 2022 | DOI:10.1163/9789004499676_006

FIGURE 4.1 Air pump by Jan van Musschenbroek, Leyden, 1698
UNIVERSITY MUSEUM GRONINGEN, INV. NO. 2020/007; PHOTO BY
S.L. ACKERMANN

instrument for the cabinet of physics: an air pump, made by the Dutch instrument maker Jan van Musschenbroek (1687–1748) (Fig. 4.1).[2] In arranging for this purchase, Bernoulli intended to use the air pump during his lectures to illustrate the vacuum phenomena that other scientists had recently discovered by using air pumps. In order to give more background to his lectures he thought it was important for the students to gain first hand, hands-on insights. To achieve this, he had indeed put up a small cabinet of instruments and he demonstrated these to students in the choir of the Academy's church (Fig. 4.2).[3] One of his experiments involved a vacuum flask filled with mercury, in which he discovered

---

2 University Museum Groningen (UMG), inv. no. 2020/007. See: Peter de Clercq, *At the Sign of the Oriental Lamp. The Musschenbroek Workshop in Leiden, 1660–1750*, Erasmus Publishing, Rotterdam, 1997, pp. 143–145.

3 See for example: *Series Lectionum 1697*; Special Collections University Library Groningen, SL1697.

Dₙ. JOHANNES BERNOULLI, Phil. & Med. Doct.

Matheſeos Prof. Ordin. Doctrinam Proportionum & figurarum ſimilium ex V & VI. lib. Elem. Euclid, Lectionibus publicis hora nona illuſtrare pergit & propediem abſolvet, poſtea Geographica & Aſtronomica ſecundum tam veterum quam Neotericorum mentem tractabit. In Collegiis vero privatis præter mathe-·matica etiam varia philoſophica docet , quocirca finito curſu Logico aperiet ineunte anno proximo 1698. Collegium phyſicum ubi omnia quam fieri poteſt commodiſſimè ad rigorem Geometricum demonſtrabit & ſubinde experimentis confirmabit. Cæterum in Algebraicis, Opticis, Staticis, Mechanicis cæterifque Matheſeos partibus, præprimis in novo & infolito calcu-landi genere ſeu differentialium & integralium methodo paucis cognita, quâ mira producuntur inventa, operam ſuam petituris non denegabit.

FIGURE 4.2 Excerpt from *Series Lectionum 1697*
SPECIAL COLLECTIONS UNIVERSITY LIBRARY GRONINGEN, SL1697

the phenomenon of luminescence.[4] His French publication concerning this[5] influenced the development of the electrostatic machine of Francis Hauksbee the Elder (1660–1713).[6]

Sadly, though, most people regarded his experiments and lectures as a fri-volity, not fit for performing in the church choir. Worst of all, the university dignitaries were not convinced of the importance of such experiments. Being a university of protestant denomination, the secrets of nature were not to be doubted, discussed or even researched by mortal man. A remark of Bernoulli (about the fact that a man's body is not the same this year as it was the last year) led to an unfriendly controversy between the scientist and a theology stu-dent, who accused Bernoulli of heresy by denying the Resurrection. Ultimately, Bernoulli left the university, as he was unable to conduct science research or lectures in the way he thought best.[7]

This small episode is not only telling for the importance of using instru-ments to gather deeper knowledge about the relationship between sci-ence and religion, but also of having an open mind towards another's work. Bernoulli left the university, but the Musschenbroek pump is still in the col-lections of the University Museum. Due to its fragility, it is impossible to use in

---

4 See: de Clercq, *Op. Cit.* (n. 2), pp. 122–125.
5 Johann Bernoulli, "Nouvelle maniere de rendre les baromètres lumineux", *Mémoires de l'Académie Royale des Sciences de Paris*, Annèe MDCC (1761), pp. 5–8.
6 See: Gad Freudenthal, "Early Electricity between Chemistry and Physics: The Simultaneous Itineraries of Francis Hauksbee, Samuel Wall, and Pierre Polinière", *Historical Studies in the Physical Sciences* 11, 2 (1981), pp. 203–229.
7 Klaas van Berkel, *Universiteit van het Noorden: Vier eeuwen academisch leven in Groningen*, Verloren, Hilversum, 2014, 4 vols.: v. 1, pp. 251–259.

demonstrations, but it serves as a reminder of how an opinion can get in the way of science. Therefore, it can be an interesting discussion starter in humanities studies.

Interestingly, about one hundred years later, the theology department played an important role in the spreading of science. At that time, theologians argued that learning about and understanding all aspects of nature would only enhance students' awareness of the perfection of its Creator. Under this outlook, the scientific understanding of lightning became enlightening both for appreciating the Creator, and for protecting human dwellings.

Every year, lightning strikes burned down churches and farmhouses with great economic and personal loss. Although science had proved that this damage was easily prevented by placing conductors on high buildings, lightning's destruction was widely accepted by traditional folklore as a punishment from God, and thus unavoidable. It is for this reason that ministers were called upon by the professor of theology Jacobus Albertus Uilkens (1772–1825) to use their influence and give demonstrations with physics instruments to try to convince church communities, as well as farmers, of the importance and possibility of protection against natural phenomena (Uilkens was an advocate of the physico-theological tradition, which stated that "the greatness of the Lord reveals itself in nature").[8] The ministers conducted this demonstration with an instrument that is a model of a church. A bowl, filled with gunpowder is placed in the model. When the demonstrator applies a high-voltage discharge from Leyden jars to the model church, it explodes and comes apart in pieces. Next, the demonstrator places a grounded conductor on the roof of the model, puts in a new bowl of gunpowder, and then applies the Leyden jar discharge to the grounded conductor, and the model remains intact.

The original instrument, the so-called "Donderkerkje" (thunder church) and Leyden jars which were used then, are still in use today. When I lecture to a live audience, or even on national television, I use the original wooden church model together with a later Wimshurst electrical machine – in place of the historical Leyden jar – to demonstrate how a grounded conductor prevents damage from lightning and other wonders of static electricity (without the black gunpowder, which was originally used).

---

8 Y. Botke, *Jacobus Albertus Uilkens,1772–1825: Predikant te Lellens en Eenrum, hoogleraar in de Landhuishoudkunde te Groningen* (exhibition catalogue), Groningen, Universiteitsmuseum Groningen, 1984, p. 9.

USING ORIGINAL INSTRUMENTS FROM A MUSEUM COLLECTION   69

Over the years, thousands of instruments have been collected from various university departments. We try to show them as much as possible, inside and outside the museum, and use them too. For many visitors, the University Museum is their first contact with scientific research, or the academic world, so it is important to offer special exhibitions that give a broader insight in the history of conducting science. This has to be on a basic level; but for groups of students from secondary schools, we run special programmes (see *Case 3* below).

3   Societies

A group of students founded the Groningen Physics Society in 1801. As with most physics societies in the Netherlands, their aim was to enhance scientific knowledge amongst their members and later a general public. Sibrandus Stratingh (1785–1841), one of the founding students, later became a professor, instrument maker and businessman/entrepreneur. Best known for his electric carriage, Stratingh also built a steam wagon, and invented an internal combustion engine as early as 1840.

The society replicated experiments done by contemporary scientists. In order to do so, its members acquired several instruments: electrostatic machines, air pumps, electromagnets, and a planetarium. Around 1880 the instruments collections were no longer of use and brought to auction. The University bought most of them; others went to secondary schools, including the newly formed Hogere Burger School. Like the original Physics Society, such new institutions emphasised more on physics and mathematics, rather than classical studies, which entitled students to go to university.

Through the years, some of these instruments have now found their way to our museum. Some are in use the way the Physics Society intended: to aid in lecturing and performing experiments that give basic insights in physical phenomena.

In the next five cases, I will present different demonstrations and/or presentations in which instruments from our collections are used. The instruments mentioned come from different collections, for example, the former Physics Society, or from laboratories dating back to the early twentieth century. The different cases also address different educational purposes. For instance, *Cases 2* and *3* highlight the work of scientists who strive for a better world, whilst *Case 1* tries to give insight in scientific content, and *Case 4* shows that, through time, instruments change, but the research stays the same.

4   Case 1: The Bernoulli Brachistochrone Solution

In 1995, the University Museum hosted an exhibition on the works of the Bernoulli family.[9] Johann Bernoulli, already mentioned above, was appointed professor in 1695. He was a member of a larger family of mathematicians, known throughout Europe. Daniel Bernoulli (1700–1782), son of Johann, born in Groningen, is famous for his work on fluid mechanics. Explanation of this played an important role in the exhibition, because you are confronted with fluid mechanics everyday (why does the shower curtain always cling to one's legs?) and secondly because it is possible to explain these phenomena using interactive displays, which makes it accessible to a broad audience.

The exhibition engaged the audience in explaining science in former times with demonstrations of mathematical and physics phenomena. As far as the mathematical content was concerned, this was set up in a game hall style, as it mainly revolved around probability – the work of Johann's elder brother Jakob Bernoulli (1655–1705) – and had no connection to old instruments. For the fluid mechanics section, we constructed several devices that demonstrated the function of chimneys, airplane wings and perfume sprayers. And yes, a life-size shower cubicle!

The exhibit presented Johann Bernoulli's famous challenge to the scientists of Europe. During his stay in Groningen, Johann solved the brachistochrone problem.[10] Without disclosing his own calculations to the mathematical community, Bernoulli wrote out the problem and challenged European scientists to solve it. Isaac Newton (1643–1727) was among those who produced a solution.[11] The *Physices elementa mathematica* by Willem Jacob 's Gravesande (1688–1742) provides drawings and calculations referring to a physical example of the brachistochrone (Fig. 4.3).[12]

Our museum's collection holds an eighteenth-century brachistochrone demonstration model. The university bought this from the estate of Professor

---

9   *Bernoulli & Zn, een complexe familie (Bernoulli & Sons: A Complex Family)*, University Museum Groningen, June 3–October 29, 1995.

10 J.A. van Maanen, *Een complexe grootheid, leven en werk van Johann Bernoulli 1667–1748*, Epsilon Uitgaven, Utrecht, 1995, pp. 77–92, and exp. p. 86. The brachistochrone problem tries to solve the following question: launching two balls, each on a given frictionless track, which is the fastest between two given points, and what formula describes the fastest curve. For clarity, in most demonstration models, one track is straight and one curved.

11 Newton's contribution was anonymous, but Bernoulli recognized the style and stated: "The lion can be recognized by his paw". See *ibid.*, p. 88.

12 Willem Jacob 's Gravesande, *Physices elementa mathematica experimentis confirmata*, Langerak & Verbeek, Leiden, 1742, 2 vols: v. 1, pp. 119–130 and pl. 17. The 1742 edition of this work is in the museum's collection: UMG, inv. no. 2020/012.

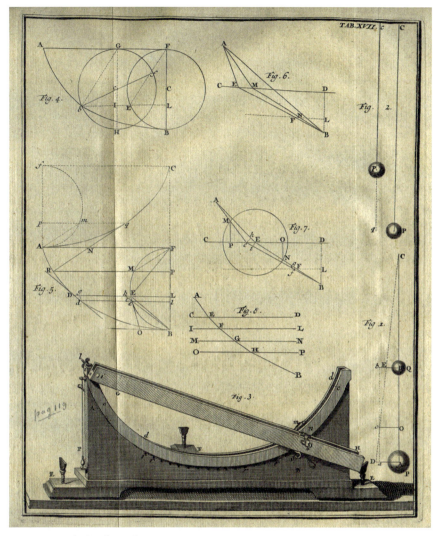

FIGURE 4.3 The brachistochrone
WILLEM JACOB 'S GRAVESANDE, *PHYSICES ELEMENTA MATHEMATICA EXPERIMENTIS CONFIRMATA*, LANGERAK & VERBEEK, LEIDEN, 1742, V. 1, PL. 17

Brugman. He held a private collection that he used during his classes, which was auctioned after his death.

In the earlier years of the museum, which was founded in the 1930s, the museum assistant (he was neither a curator nor a collections specialist) used this model to demonstrate to visitors. In the case of the 1995 exhibition, it was unsuitable, because by this time it was recognised that museum visitors should not handle an object of this age. In addition, because the exhibition was to be presented at different venues, in unguarded spaces, it had to be stand-alone, without balls flying around (which will inevitably be the case with a group of children) and the speed- and time-difference between the two tracks should be clearly visible. We succeeded in reaching all the goals with a computer- controlled golf ball loaded device, with a size of 100 × 200 cm. The visitor could start the experiment with just a push on a button and in the meantime, a computer screen presented information in words and graphics. It performed faultlessly during the months of the exhibition and at three subsequent venues.

Although this setup did what it had to do, it turned out to be less suitable for educational purposes in the normal museum setting or as a pop-up instrument. It was too big and took a considerable time to set up. Armed with this experience, we had a replica made, precisely copied from the instrument in our collection. Later, a new version was made from acrylic and looked more like a modern demonstration instrument. Finally, we made a copy, sized 1.5 × 3 m, to demonstrate at the visit of H.M. the Queen of the Netherlands to Groningen during her birthday celebrations on April 30, 2004. It could not be missed; it was made out of fluorescent orange (in honour of the royal family colour) acrylic and stood about 2 m tall! It attracted a lot of interest; hundreds of people passed that day and stopped to listen to our explanation.

While the historical background of the brachistochrone was purely mathematical, and researched by the greatest minds of their time, it is now an instrument to entertain the audience, but also to give an understanding about mathematical and physical problems (that is, the influence of gravity on the acceleration along different curves). Hence the different versions, each serving its own purpose and audience.

The problem with a perfect copy, such as the first of the three versions of the brachistochrone apparatus that we commissioned, is that it is hard for the visitor to distinguish between old and new. Visitors may be so confused that they think the museum uses real artefacts for hands on demonstrations. In itself, that should not be a problem, as long as the right person handles them.

5 Case 2: The Smoked-Drum Kymograph

From 4 to 14 April 2013, Imperial College London hosted the exhibition *Strictly Science*. This was to celebrate the centenary of the (British) Medical Research Council.[13] As such, this exhibit had no connections with the University of

---

13   The free interactive exhibition *Strictly Science: Keeping One Step Ahead* invited visitors to engage in the past, present and future of medical research. The public watched scientists

Groningen, nor the University Museum. Our cooperation started with a question on the RETE forum: "I'm organising a public exhibition about medical research past, present and future to commemorate 100 years of the Medical Research Council in 2013 at Imperial College London from 4th–14th April. The exhibitions team is looking for a kymograph (preferably a late 19th century/early 20th century model) to help demonstrate the research methodology used by Sir Henry Dale, who discovered acetylcholine and demonstrated its antagonistic effects on cardiac and skeletal muscle".[14]

The organisers needed an early twentieth-century kymograph, for use in the recreation of a laboratory from about 1913. No institutions in the UK were willing to lend this museological object for a set up in which it would be really used. We offered to loan one of our historical kymographs from the collections of the former laboratory for psychology to this exhibit. I prepared to demonstrate it live by redoing the original medical experiments conducted by Henry Hallett Dale (1875–1968).[15]

When I studied the experiment as performed by Dale, in which he used animal organs, injected them with chemicals and recorded the reaction of said organs, some problems arose for reconstructing it as a public demonstration today. He made use of rabbit legs, frogs' hearts, chemicals and syringes and needles; not the things that health and safety officers are happy about in the public domain! The kymograph (Fig. 4.4) is an instrument that draws a graphical representation of stimuli over time. It uses smoked paper on a rotating drum as a recording device. Smoking paper, in order to cover it in an equal layer of soot to be written on by a scribe point, is an art, which is long lost. Operating the kymograph for the full ten-day UK event would require a large and steady supply of smoked paper, and this was considered unpractical. Therefore, decisions about an alternative had to be found.

Normally, when preparing experiments like this, people sit around the table and discuss the problems. In this case, due to our being a great distance apart, all discussion was by email, telephone and Skype. Haberdashery, a London-based exhibitions builder, constructed a faux rabbit leg; this was mounted in a prepared metal jar. By coupling the faux leg to an Arduino based

---

conduct century-old experiments and interact with contemporary neuro-technologists investigating the brain. This was a rare opportunity for the public to learn about the enormous contribution taxpayer revenue has made, and continues to make to medical practice in Britain and worldwide.

14   See RETE archive: https://web.maillist.ox.ac.uk/ox/arc/rete, keyword "kymograph", message dated November 29. 2012 (accessed Nov. 17, 2020).

15   Sir Henry Hallett Dale (1875–1968) was an English pharmacologist and physiologist. For his study of acetylcholine as agent in the chemical transmission of nerve pulses (neurotransmission) he shared the 1936 Nobel Prize in Physiology or Medicine with Otto Loewi.

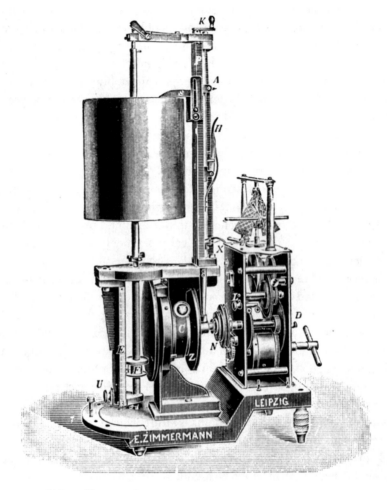

FIGURE 4.4 Universal kymograph
 EDUARD ZIMMERMANN, *WISSENSCHAFTLISCHE APPARATE: LISTE 50*, HEINE & CO., LEIPZIG, 1928, P. 188; UNIVERSITY MUSEUM GRONINGEN, INV. NO. UMG DEPOT H65

system, it could be contracted with a remote switch. The leg was mechanically coupled to the recording levers of the kymograph. The principle worked all right, but during build up and the run of the exhibition several problems arose. After the Arduino began to lead a life of its own, with the potential to harm the fragile levers of the historical instrument, we decided not to use it. In the end, we worked the system with a few cords and a pulley.

Instead of marking smoked paper we ran prints, copied from originals, on the kymograph as it rotated. During the demonstration I explained the idea to the audience.

Together with the two other experiments, performed by British educators, we presented a programme from 11:00 AM to 5:00 PM, in front of a broad audience of all ages. The most rewarding part was the interest and patience of the people who attended the exhibition. As an extra bonus, I was able to speak to several people who worked with kymographs like ours in former days. This is where you find that public engagement should be a key element in every museum. Working with a dedicated and knowledgeable team of demonstrators (all of them had a PhD) and organisers, and interacting with the audience are the important ingredients for a successful demonstration.

6 Case 3: Stratingh's Electrical Machines

One of the elements in the final exam for Dutch secondary school pupils is doing a school research project.[16] This can be a paper, a construction, a scale model, almost anything, but must have a connection to the chosen exam profile (health/environment, physics/engineering, economic or cultural studies related). To support this, the so-called "Scholierenacademie" of the university organises a week called "Onderzoeksdagen" (Research Days) of activities to support these pupils. The role of the museum in this is to demonstrate the breadth of academic studies and bring physical context to the lectures during the week. This contribution is in co-operation with other heritage departments, such as the University Libraries Special Collections, which shows old manuscripts, and the University Hospital, which uses our anatomical collections.

For this programme, we put up small demonstrations with historical apparatus used in experimental psychology, biology and chemistry/physics. A central element in the last example is electromagnetism/electricity, coupled to green energy and sustainability. In this, we show old electric motors and generators, describe the development of technique and society, and try to let the pupils think of solutions for current problems, that is, how to cope with the growing demand for electricity.

The work of Stratingh stars in these presentations. We show his original electric motor and carriage, which are both too fragile (and unique) to be demonstrated. Therefore, we had both replicated. These working copies are excellent instruments to show how an electric motor works and how (electro) magnetism can be put to use.

---

16   See: www.rug.nl/society-business/scholierenacademie/scholieren/pws-hulp/ (accessed Dec. 30, 2020).

FIGURE 4.5 *On the left*, electric motor by Stratingh and Becker, 1835; *on the right*, replica by A. Stoelwinder, 2009
UNIVERSITY MUSEUM GRONINGEN, INV. NO. 1986/052 AND AUTHOR'S PRIVATE COLLECTION; PHOTO BY S.L. ACKERMANN

Before his work on electric propulsion, Stratingh built a steam carriage. Unfortunately, this no longer exists; the only things left are some contemporary sketches. To bring to life its role in the story and the development of transport, the museum is also commissioning a scale model of the steam carriage. Based on the old drawings it can only be a reconstruction, not a replica. Sadly, a life size model (where museum visitors could ride in the steam carriage!) is beyond our means …

The electric motor (Fig. 4.5) was presented live during the *XXXVIII Symposium of the SIC* in Havana. In addition, a video offered a preview of the internal combustion engine proposed by Stratingh around 1840. During his lifetime, it was not yet technically possible to construct this engine, but the original drawing gave an indication as to its construction. In 2018 Anton Stoelwinder, from Gorredijk, the Netherlands, who also built the other replicas in use in our museum, made a reconstruction. Technically, this engine runs on the combustion of oxyhydrogen (H_2+O_2) also known as Knallgas, but can be regarded as a hydrogen engine. It uses all the components that define an internal combustion engine (spark plugs, timed ignition, valves and four-stroke cycle).

With these three instruments, it is possible to show the full scope of the development of a scientific principle and its impact on society and climate.

7   Case 4: Instruments for Experimental Psychology

As mentioned above, instruments used in experimental psychology were also demonstrated during the "Onderzoeksdagen". These come from the former laboratory for experimental psychology, which was founded in 1892 by Gerard Heymans (1857–1930). He bought instruments mainly from the catalogue of the German manufacturer Zimmermann in Leipzig. All these instruments have survived to this day, thanks to the good care of subsequent laboratory directors. Nowadays they form one of the most important collections in the Museum. Thanks to the quality of construction, they have survived well and are in excellent condition. (The large kymograph, used during the aforementioned London event, is from this collection).

Why use nineteenth-century instruments when there are electronic devices that are easier to set up and work with? Take for instance a memory experiment (Fig. 4.6). Nowadays you would use a computer and screen with a keyboard of some sort and/or a microphone. Press start and things just happen. However, when you get back to the early twentieth century, you would need a clock, a timer, an instrument to provide impulses, an acoustic switch, a recorder and miscellaneous connectors, wires and switches. You have to build and test the setup, which gives you much more insight into what is actually happening. During the experiment, you have to adjust instruments and perhaps wind up clocks, so you are more engaged in the process. Because of the transparency of the materials, the person to be tested can see what is happening, which, hopefully, gives more trust in the process.

The memory experiment also shows that, although the instruments change, the line of research stays more or less the same. When Heymans started his research on weight perception (the so-called physical inhibition),[17] he collected large quantities of data and applied statistical analysis on this.[18] Statistics is still one of the most important parts of the curriculum today. In addition, the principal basics of distracting influences in traffic were discovered by him. He did this with a simple siren and rotating colour discs; in around the year 1990 the Psychological Institute used a driving simulator built around a complete car, using rugged computers made to military specifications.

---

17   The pendulum balance used for this research is in the collections of the University Museum, inv. no. 1962/23,23.

18   Kars Dekker, *Gerard Heymans, Grondlegger van de Nederlandse psychologie*, WB Uitgeverij, Groningen, 2011, pp. 304–315.

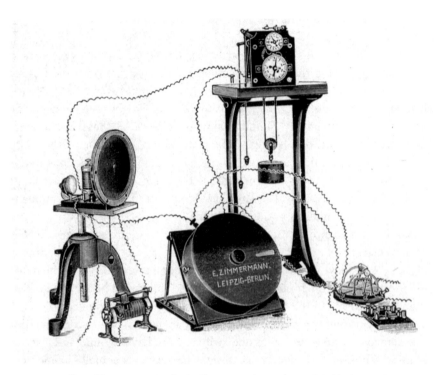

FIGURE 4.6 Ranschburg apparatus for testing perception and associated instruments
EDUARD ZIMMERMANN, *WISSENSCHAFTLISCHE APPARATE: LISTE 50*, HEINE & CO., LEIPZIG, 1928, P. 64; UNIVERSITY MUSEUM GRONINGEN, INV. NO. UMG DEPOT H65

Even the names of the instruments stay the same. The picture of the memory experiment shows (Fig. 4.6, on the left) a "Schallschlüssel", by the German firm Zimmermann. Ninety years later, the electronics workshop of the institute made an electronic version of this and called it "Voice Key", which translates as "Schallschlüssel". However, they were unaware of this fact, as they were not regular visitors of the museum …

When performing all this to schoolchildren, they can get an insight into what is involved in performing experiments, and that science not only involves sitting behind a desk, but also is a hands-on activity. In an ideal world, the whole setup should be designed and built by a group of students. Nowadays, when school physics practical classes mainly involve redoing pre-defined experiments, students' participation is mainly a matter of reading a display and putting results in a graph. Building more complex mechanical and electrical setups with a historical context is impossible in the current secondary school curriculum, but has potential for a project during school visits to the museum.

Making old experiments work is also a challenge for the museum staff. Early twentieth-century electrical devices usually worked on battery power, because mains electricity was not available. To emulate battery power electricity with today's equipment involves getting power supplies that provide low voltage and high current output, which, in turn, demands wiring that is not currently common. In addition, health and safety requirements have to be taken in account. Therefore, before you can even start a demonstration, the staff has to have gained considerable knowledge about the experiment itself and the techniques that make it work.

The same goes for many other disciplines. Seeing old documents, and being able to touch them (under controlled circumstances) makes history tangible and brings it closer to you. Holding a meteorite or fossil in your hand, in the knowledge that this piece of rock is older than the earth or shows an organism which walked on that earth long before human presence, can be an experience that suddenly makes you aware of what you want to study after secondary school.

For medical classes, professors come to the museum asking to borrow human preparations to use as study material. They have come to the conclusion that computer-generated material just will not do, even if these show much more detail than somewhat discoloured one hundred-year-old fluid preparations of the brain, or diseases.

A real three-dimensional view from something you actually can handle (albeit within museum constraints) can give a better topographical reference. Apart from that, it can be an experience that (hopefully) lasts longer then a computer-generated image.

All the examples in these cases have a common value: they show where science is coming from, which makes it easier to understand where you can or have to go, and that different disciplines are interconnected. That is the message we want to give to the secondary school pupils, but also to the public, when giving lectures and demonstrations.

8 Case 5: Future Plans

In 2021 the museum hopes to open a new permanent exhibition about the history of science in general and at our university in particular. The former permanent exhibition opened in 2005 and had some small changes over the years. However, the only information about the instruments were small texts on an object level, stating name, manufacturer and year, and some general information about the contents of the showcase in longer texts. Due to adverse circumstances, the handy printed guide to the collections never emerged.

FIGURE 4.7 Ribokov cage, unknown maker, third quarter of twentieth century
COLLECTIONS ARCHIEF EN DOCUMENTATIECENTRUM NEDERLANDSE
GEDRAGSWETENSCHAPPEN – ADNG, INV. NO. A40; PHOTO BY
S.L. ACKERMANN

It is therefore time to reopen with a much more activating and interactive exhibit. The history of science will still be the leading theme, but will allow for the fact that history never stops, so the new exhibit will look into developments after tomorrow. This will be done by interactive devices, videos and demonstrations. These demonstrations will be part of guided tours including hands-on activities for visitors. Learning from experiences as described earlier, we will use replicas, but also, if possible, original instruments. We plan to allow visitors to interact with historical instruments, especially from the collections on experimental and occupational psychology; such objects are either plentiful, or have a construction quality that makes them virtually indestructible (see, for example, the Ribokov Cage in Fig. 4.7).[19] An original Leybold's

---

19   The Ribokov Cage is an instrument to study hand-eye coordination. It is a tabletop round cage with a curved obstacle course inside. The person who is the subject in this test has to work two rings from the bottom to the top, using two sticks. The instrument

USING ORIGINAL INSTRUMENTS FROM A MUSEUM COLLECTION 81

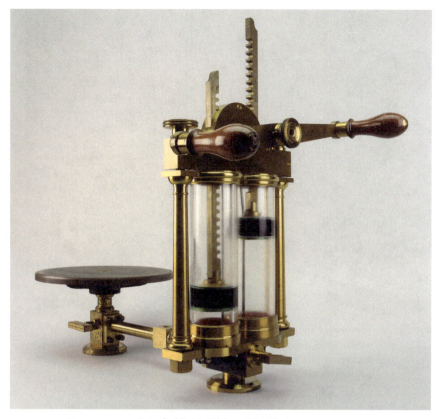

FIGURE 4.8 Vacuum pump, E. Leybold's Nachfolger Cöln-Rhein, 1905
UNIVERSITY MUSEUM GRONINGEN, INV. NO. 2019/45,048; PHOTO BY
S.L. ACKERMANN

Nachfolger vacuum pump, which was originally built for demonstrations, is also a candidate for use in a controlled setting (Fig. 4.8).[20]

As electromagnetism plays a very important role in science and in society, this will be one of the focal points. Through electrostatics, the work of Stratingh, and early X-ray discoveries, we will reach the latest research on particle physics. The idea is to reconstruct a workspace/laboratory, to create an engaging environment, not unlike the "Lorentz Lab" at Teylers Museum in Haarlem,[21] for the audience to experience and discover. As with every museum, the target

  comes from the collections of the Archief en Documentatiecentrum Nederlandse Gedragswetenschappen – ADNG (Dutch Archive for the Behavioral Sciences), inv. no. A40.
20 University Museum Groningen, inv. no. 2019/45,048.
21 See the chapter by Trienke van der Spek in this volume: *infra*, pp. 243–262.

FIGURE 4.9 Experiment for testing memory and perception; setup used in 1980 in an exhibition celebrating the seventeenth birthday of the Psychological Institute.
PHOTO BY THE PHOTOGRAPHIC SERVICES DEPARTMENT OF THE UNIVERSITY OF GRONINGEN

group is an audience from 8 to 80 years of age, but focus will be on secondary schools and university students.

Besides displaying scientific (physics) instruments, there will also be a place for issues and instruments relating to the social sciences and the humanities, and biology and medicine.

One future project cannot be left unmentioned: a proposal from a Belgian PhD student, who wants to research ways to measure and counteract the effects of migraine, based on the works of Hildegard von Bingen (1098–1179) and experiments done in the 1930s, using, amongst others, a kymograph and ancillaries from our collection.[22]

9   Conclusions

Over time, much has changed in the way we regard using museum artefacts as demonstration material. In the 1950s the museum assistant regularly used old

---

22   See: J.N. Blau, "Harold G Wolff: the man and his migraine", *Cephalalgia* 24 (2004), pp. 215–222.

USING ORIGINAL INSTRUMENTS FROM A MUSEUM COLLECTION   83

and original instruments as demonstration material.[23] In more recent years, museum ethics put a stop to this, but nowadays showing instruments outside their showcases is making a comeback. There is a line to be drawn of course, and that is where replicas enter the stage. Although expensive to make, and not always easy to use, they can shed light on older techniques and the way man had to work. Besides, they are attractive to see and can draw the audience into a story about certain research, a person, or social matters. As stated earlier, it can be confusing for the museum visitor when a replica, almost indistinguishable from the original, is used. Therefore, it is best to show and use these in a controlled setting, explaining why this is done. Even a simple instrument, like the original small Stratingh electric motor is too delicate to be handled by the audience.

As the experiences from the co-operation with the Medical Research Council proved, a laboratory setting works very well, but trained staff (not only knowledgeable about the instrument, but also about its place in history, and history in general, and therefore able to tell a story in its correct perspective) is the key to success. To perform something similar on a daily basis in a (small) museum is almost impossible. It would take a pool of professional science educators, which is beyond the financial means of most institutions, except the largest ones. Therefore, it will depend on the willingness and the abilities of the daily staff to perform the act, on a maximum of a weekly basis. In this case, we are lucky that, during the last few years, the museum has merged with the Universities Centre for Cultural Students Activities (USVA), the Scholierenacademie and Studium Generale (which organises multidisciplinary activities in the field of science, culture and society), in a new "Centre for Public Engagement". That name says it all and this combination brings new possibilities and a fresh view from outside the museum on how to reach the public.

However, the most important objective for a science museum is that instruments, be they real or replicas, are always at the centre of attention, be it during a lecture, a presentation or a demonstration in front of the audience. An audience is amazed when they see how in 1835 someone built an electric car that still is able to be driven (albeit replicated, but strictly following its original design), and is intrigued by a simple rotating drum with writing pens, which stood at the base of Nobel Prize awarded research. It is rewarding when visitors leave the lecture hall, or demonstration laboratory, saying to each other: it was a difficult matter, but now I understand how it works.

---

23   See, for example, the setup used in 1980 in an exhibition celebrating the seventeenth birthday of the Psychological Institute (Fig. 4.9); *Gerard Heymans, 1857–1930. Filosoof en psycholoog. Leven en werk* (exhibition catalogue), Universiteitmuseum Academiegebouw, Groningen, 1980. The experiment is discussed in *Case 4* above.

CHAPTER 5

The Collections of Scientific Instruments of the Faculty of Sciences of Rennes

A Tool for School Education and for the Training of Students and Teachers

Julie Priser and Dominique Bernard | ORCID: 0000-0001-6909-6651; 0000-0001-9011-034X

1 Introduction

In this chapter we present the approach that we have been carrying out for fifteen years at the University of Rennes 1, France, to develop the use of historical scientific instruments in the training of students and teachers.[1] After detailing the collection of instruments of the Faculty of Sciences, we will present the promotional activities of these collections that we have developed for schoolchildren and a wider audience. The reconstruction of the historical experiments of radioactivity measurements by Pierre and Marie Curie is discussed. We present our various interactions with research laboratories, university libraries, institutes, high schools or colleges in the Brittany region.

2 The Collections of Scientific Instruments of the Faculty of Sciences of Rennes

The University of Rennes 1 is located in the Brittany region, in the west of France.[2] From the year 2000, the University collection was gradually reconstructed after the initial approach of two physics teachers from the University of Rennes 1: Dominique Bernard and Jean-Paul Taché.[3] The collection includes

---

1   This chapter is an elaboration of the paper presented at the *XXXIX Symposium of the Scientific Instrument Commission*, London, UK, 14–18 September, 2020.

2   See the University internet website: www.univ-rennes1.fr (accessed 5 Oct. 2020).

3   On the collection see: D. Bernard, W. Tobin, A. Canard, J.-P. Taché, "The Physics Instrument Collection at the University of Rennes 1", *Scientific Instrument Society Bulletin* 111 (2011), pp. 34–39; D. Bernard, *Un trésor scientifique redécouvert – Les collections d'instruments scientifiques de la Faculté des Sciences de Rennes (1840–1900)*, Rennes en Sciences, Rennes, 2018.

© JULIE PRISER AND DOMINIQUE BERNARD, 2022 | DOI:10.1163/9789004499676_007

THE COLLECTIONS OF SCIENTIFIC INSTRUMENTS IN RENNES   85

around a thousand objects from the nineteenth century, to which are added several thousand items from the twentieth century. All scientific disciplines are present but, more particularly: physics, chemistry, electronics and computer science. Some very rare instruments are worth a look: a large tuning fork by Rudolph Koenig (1832–1901), a gyroscope by Léon Foucault (1819–1868), an electromagnet by Pierre-Ernest Weiss (1865–1940) and a telescope by James Short (1710–1768) dated 1740.[4]

The conservation area consists of a gallery located in the basement of a teaching building and several storage reserves scattered around various locations on the Beaulieu Scientific Campus. An inventory and promotion programme called PATSTEC – Patrimoine Scientifique et Technique Contemporain (Contemporary Scientific and Technical Heritage) has been set up in collaboration with the Conservatoire National des Arts et Métiers (CNAM) – Museum des Arts et Métiers. In this national programme, the University of Rennes 1 is a pilot for the Brittany region.[5]

3   The Training of Students and the Use of Scientific Instruments

3.1   *Internships for Students on Historic Scientific Heritage*
About forty students took part in internships on historic scientific heritage before 2020. They were able to develop their knowledge in the history of science, carry out bibliographic work, compare old experimental results with the latest experimental techniques in research laboratories, and also present these devices and their stories to the public. Doing these projects prepares students meaningfully for their future profession of teacher or researcher. Here are a few examples of this work. Some have given rise to publications and communications, in particular to the meetings of the Scientific Instrument Commission (SIC):

– Nathalie Rozé, physicist and musician, carried out different experiments using the acoustic instruments of Rudolph Koenig (Fig. 5.1). Her internship was a great way for her to develop an experimental practice, and work on her communication with middle and high school children. With her colleague Jérémy Thouin, she put back into service a Heron's fountain, dating

---

4   *Ibid.*, pp. 22–23 (Short's telescope), 84–87 (Koenig's large tuning fork), 174–181 (Foucault's gyroscope) and 208–213 (Weiss' electromagnet).
5   The PATSTEC mission is presented in detail on the website www.patstec.fr (accessed 5 Oct. 2020).

FIGURE 5.1 Internship students Nathalie Rozé and Jérémy Thouin operate old devices: a double Helmholtz siren; University of Rennes 1, 2005

from 1840, which ensures for us an undisputed success each time we present it in operation.[6]

– The discovery in our collections of a Violle actinometer, made by Ducretet in 1905, intrigued us. This very rare device is in the form of a metal sphere with a circular opening and cooling circuits. Jules Violle (1841–1923) had carried out very fine experiments around 1875 to determine the temperature of the surface of the Sun. As part of her master's degree in physics, Olympe Jouet put this actinometer back into operation, and then carried out radiation measurements using the results of Violle and current radiation theories. She determined a temperature of the surface of the Sun very close to that currently accepted, 5800 °K.[7]

---

6 A. Faisant, S. Bourles and D. Bernard, "Historical Instruments as Tools to Develop the Curiosity of Young People: The Example of Heron's Fountain", *XXVII Symposium of Scientific Instrument Commission*, Lisbon, Portugal, 16–21 September 2008. A short video was made by Rozé and Thouin on this Heron's fountain: www.youtube.com/watch?v=SloxEcjUOtg&list=UU9mR5zC2Qg9ElnddpoKIfSw (accessed 5 Oct. 2020).

7 O. Jouet, A. Amon, D. Bernard, "Jules Violle's Actinometer: A Simple Instrument to Deduce the Temperature of the Surface of the Sun", *Scientific Instrument Society Bulletin* 112 (2012), pp. 28–31.

THE COLLECTIONS OF SCIENTIFIC INSTRUMENTS IN RENNES 87

3.2   *Internships in Conjunction with a Research Laboratory*
A second type of research internship involves the inventory of other devices in conjunction with the Brittany programme PATSTEC. The overall objectives of this body are to create a relationship between research laboratory and scientific heritage and to recover, safeguard and inventory scientific devices from the 1950s. These internship students work both with the heritage team and with researchers in a laboratory. They are guided by us on the heritage part of each project. As regards the science, they consult with researchers from the laboratory to which they are attached. Here are some examples:

− Study of a Curie scale from 1910 (Société Centrale de Produits Chimiques) by a chemistry student, Julien Le Clanche, with work in the Glass and Ceramics Laboratory of Rennes.
− Recovery of lasers in a laboratory by Mickael Goble, master's degree in physics student. This project involved preparing inventories, writing files for the PATSTEC database, carrying out experiments with researchers from the laboratory of Laser Physics of the University of Rennes 1.
− Study of historical spectroscopes (Duboscq prism and Thollon spectroscopes) by Valentin Hervault, another student physicist. He also spotted and inventoried various spectroscopes from the 1950s to 1970s, and used some of them. As part of the internship hosted for some time at the Laboratory of Physics of Infra-Red Spectroscopy of the University of Rennes 1, he participated with scientists of that laboratory in spectroscopy experiments. Working with us at the same time, he carried out successful operations on nineteenth-century devices.

3.3   *Reconstitution of the History of a Discipline, a Laboratory*
***   or a Discovery***
In response to the enquiries that we frequently receive from other researchers or laboratories, internships can be directly linked to a historical approach to the disciplines or to the creation of laboratories. Here are a few examples:

− *When a physics student becomes a science historian.* It is interesting to describe Axel Petit's career in his approach that combines the promotion of scientific heritage, teaching and research. First as a physics student at our university, Petit did a Master 1 internship with our "Culture and Heritage" team, and undertook original work on the history of physics and physicists at the Faculty of Science in Rennes from 1840 to 1939. This historical research was accompanied by some experiments on a Fabry-Buisson micro-photometer, a very rare device made by Jobin around 1900 and kept in our gallery of scientific instruments.[8] In a continuation of this work, Petit

---

8   A. Petit, D. Bernard, "La Physique et les Physiciens à la Faculté des Sciences de Rennes de 1840 à 1939", *Internship report of Rennes 1's University* (2010), www.academia.edu/24108652/

has decided to leave Rennes and pursue this path at the University of Nantes for a Master 2 in history of science and technology at the Centre François Viète. Here he wrote a dissertation about the dean, Georges Moreau, physicist at the Faculty of Sciences of Rennes.[9] This work led to his doctoral thesis in the history of epistemology of science and technology.[10]

– *History of the discovery of new glasses in a chemistry laboratory.* Another interesting sector in chemistry is that of the history of the discovery of new types of materials: fluoride glasses which established the reputation of a solid-state chemistry laboratory in Rennes, the Verres et Céramiques Laboratory.[11] In 1975, in this laboratory, researchers discovered new glassy compounds based on fluorides. These discoveries were made with completely artisanal equipment: sealed nickel tubes, work in a "dry box", melting in an oven, or pouring in artisanal moulds designed in the laboratory. Consulting with researchers from the laboratory who are still active, Bérengère Frize, a chemistry student, was able to reconstruct the story of this discovery, find objects and devices, inventory them and interview the researchers. By interacting with the original researchers, she was able to describe how this discovery impacted and transformed the orientation of the laboratory.[12]

– *History of crystallochemistry in Rennes.* The faculty of sciences of Rennes has a long tradition of research in X-ray crystallography thanks to two laboratories created by Jean Meinnel and Daniel Grandjean (in the 1970s). Our collections of "contemporary heritage" are enriched by numerous devices from these laboratories. An internship project around the history of crystallochemistry in Rennes was proposed to Younès Bareha for his master's degree in chemistry internship. He spent a period in the laboratory, inventoried many devices such as diffraction chambers or X-ray tubes, and

---

La_physique_et_les_physiciens_%C3%Ao_la_Facult%C3%A9_des_Sciences_de_Rennes _de_1840_%C3%Ao_1939 (accessed 5 Oct. 2020).

9   A. Petit, "Georges Moreau (1868–1935), administrer, enseigner, chercher à la faculté des sciences de Rennes", *Cahiers François Viète* Série II, 4 (2011), pp. 1–31; also available at: https://cfv.univ-nantes.fr/cahiers-francois-viete-serie-ii-n-4-1254850.kjsp?RH= 1405598162629 (accessed 5 Oct. 2020).

10   A. Petit, *L'Histoire du concept d'ion au XIXe siècle*, PhD dissertation, University of Nantes – Centre François Viète, 2013.

11   See: https://iscr.univ-rennes1.fr/glasses-and-ceramics-vc (accessed Oct. 5, 2020).

12   B. Gallon and D. Bernard, "Histoire de la Chimie À la Faculté des Sciences de Rennes de 1840 à 1966", *Internship report of Rennes 1's University* (2010), www.academia.edu/ 24109101/Histoire_de_la_Chimie_%C3%80_la_Facult%C3%A9_des_Sciences_de_Rennes (accessed 5 Oct. 2020).

wrote a historical summary. His work contributed to an exhibition for the International Year of Crystallography in 2014.[13]

– *History of the Chemistry laboratories of the Faculty of Sciences of Rennes.* Although Louis Joubin documented the nineteenth-century Rennes faculty of science, his volume does not discuss the subject of chemistry in detail.[14] With two master's students in chemistry, Benjamin Gallon and Julie Priser, we set out to write a history of chemistry in Rennes from 1840 to 1980. These students carried out archival research, interviewed researchers, and visited laboratories. In addition to raising the awareness of chemists today about the conservation of instruments, these contacts provided an opportunity to enrich the collection of instruments and to write inventory sheets for databases.[15]

– *2020, Ampère Year – Bicentenary of his discoveries.*[16] The year 2020 marked the 200th anniversary of the fundamental discoveries made by André-Marie Ampère (1775–1836). We suggested to Graziella Guy, a third year student in physics and chemistry, to do an internship working on the collections of scientific instruments, for work of electrodynamics with a specific focus on those relating to the work of Ampère and other physicists of the time (Ørsted, Arago, Biot, Faraday, and others). Guy operated some of the original nineteenth- and also twentieth-centuries devices made by Eugène Ducretet (1844–1915) and Emile Deyrolle (1838–1917),[17] like the Ampère table, the Ampère table modified by Bertin, and the Obellianne device. Unfortunately, the Covid-19 epidemic severely disrupted the celebration year, and activities were halted.

---

13   During the International Year of Crystallography 2014, several exhibitions took place in Rennes and were conducted by Julie Priser, Audrey Chambet, Gaëlle Richard, Jean Plaine and Marie-Aude Lefeuvre. All such exhibitions are summed up at: https://rennes.udppc .asso.fr/IMG/pdf/DossierPresseCristalloLycee.pdf (accessed 24 Apr. 2021).

14   Louis Joubin, *Histoire de La Faculté des sciences de Rennes*, Rennes, Imprimerie Fr. Simon, 1900.

15   B. Gallon, J. Priser, A. Perrin, C. Perrin, D. Bernard, A. Chambet, "L'histoire de la chimie à la Faculté des sciences de Rennes", *Les Cahiers de Rennes en Sciences* … (2021), pp ……, in press.

16   See: https://ampere2020.fr/ (accessed 5 Oct. 2020).

17   Eugène Ducretet was an important French scientific instrument manufacturer at 75 rue Claude Bernard, Paris. Emile Deyrolle was a French naturalist and natural history dealer in Paris. The company – which still exists at 46 rue du Bac, Paris – also distributed scientific instruments under the firms of "Les fils d'Émile Deyrolle" and "Eurosap-Deyrolle".

4 Students and Scientific Heritage

Since the beginning of reconstructing collections of historic instruments, pioneers have worked to educate science students and others about the value of these pedagogic devices. This awareness has therefore taken several forms:

1. Organisation of collection visits, including: student organisations; welcoming new students at the start of the academic year; and presentations of experimental demonstrations.

2. Production of exhibitions. The university library of the Faculty of Sciences, located very close to the gallery of instruments, is a very suitable place to meet students. Its reception hall is suitable for presenting rare and precious objects and books, and for telling stories. This place, very popular with students, lends itself well to demonstrations. Thanks to the constant collaboration of the team of the central documentation service of Campus Beaulieu, we have been able to organise numerous exhibitions whose theme is linked to the current International Year of the United Nations, such as: *From natural history to biodiversity* (2010), *From mineralogy to crystallography* (2014), *25 centuries of light* (2015), *Minitel, the ancestor of the Internet?* (2015), *Discovering legumes* (2016), *Mendeleev's classification of chemical elements* (2019).[18]

The organisation of all these exhibitions was led by André and Christiane Perrin, two CNRS researchers who are members of the Rennes Institute of Chemistry. At each exhibition, explanatory posters, instruments and objects, books and old documents from the library are presented, as well as recently-published books. Following the inauguration of the event, guided tours of the exhibitions are offered to students and interested audiences.

Exhibitions have also been organised in a building of the University's Culture-Heritage Service called "Le Diapason".[19] It is a meeting place for students and staff that brings together sporting, musical and cultural activities. Some of the exhibitions presented there have been: *GLASS: Object of Science and Art* (2017), *Objects for pedagogy. Observation and description tools for science education* (2018).

3. The QUESACO project for students (2018–2019). In partnership with the university's Fablab, we have designed a project called QUESACO – QUEstioning Scientific COllections: from heritage to innovation. Each year, we select an instrument from our collection; it serves as a "challenge object" for a student competition. Students are invited to meet this challenge by working together

---

18   See: https://bibliotheques.univ-rennes1.fr/les-expositions-virtuelles-o (accessed 5 Oct. 2020).

19   See: https://diapason.univ-rennes1.fr (accessed 5 Oct. 2020).

in interdisciplinary teams of three to four people. The game consists of imagining and building an innovative device that improves and responds to the historical object. What the student teams produce to meet this challenge is open: modern prototype; artistic and/or scientific mediation (video, presentation, and so on); or take a new form that we have not imagined.

Although the "challenge object" has the effect of promoting the collections, the project steering committee invites an active and diverse approach from the students (bibliography, experiments, presentation of the work to a jury and to the public).

The "challenge object" that featured in the first QUESACO in 2018, was a Rudolph Koenig's sound analyser from 1870 (Fig. 5.2), an exceptional invention due to its degree of innovation and its astonishing shape. A questionnaire, carried out by the project managers, was set out on different aspects of the device: historical, technological, scientific, and possible uses. Students could also contact teacher-researchers. Communication towards the students around this event was classic (posters, websites) but a "teaser" had also been designed by the teachers to attract them.[20] A large investment of time and effort by the École Supérieure d'Ingénieurs de Rennes (ESIR) staff members, such as Olivier Ridoux and Pierre-Antoine Angelini, ensured the success of this event. Four student teams were selected and presented their results to the public.

In 2019, this project was fundamentally based on a participatory, multidisciplinary training approach, open to all. Its starting points involved objects from the collections of scientific instruments: Jules Violle's actinometer, photographic enlarger and calculating instruments. The point of arrival was an innovative competition during the organisation of an event. This event consisted of a presentation of the students' work in front of a jury and an audience during a special evening. Under the impetus of a computer science teacher at the University of Rennes 1, Olivier Ridoux, this project was organised at the ESIR within a "course innovation". Several groups of students had a whole year to prepare research work in the history of science and technology, which culminated in an exhibition open to all.

Project participants provided enthusiastic feedback. Indeed, the QUESACO project, quite original in form, allowed students to take time to work on science history, to research an idea for introducing their presentation, to work also in a theatrical environment, to improve teamwork and communication skills. This project gave a practical dimension to their university studies.

---

20   See: www.youtube.com/watch?v=WSqEGuFFy7U and https://esir.univ-rennes1.fr/actual
      ites/lesir-recompensee-au-challenge-quesaco (both accessed 5 Oct. 2020).

FIGURE 5.2 The QUESACO project with Rudolph Koenig's sound analyser; University of Rennes 1, 2017

5 Teachers and Scientific Instruments

5.1 *Scientific Collections: A Place of Reception for Teachers*

For teachers, whatever the level of classes they supervise (from elementary school to high school), the collections are always a place of welcome. Contacts are well established with their associations, such as the *Union des Professeurs de Physique et de Chimie* (UDPPC – Union of Professors of Physics and Chemistry), or directly with their institutions. Special consultation visits are organised for them, in preparation for visits by their classes.

For certain visits or exhibitions, we welcome scientific experts from our network such as William Tobin and Alain Faisant, who study Léon Foucault's instruments (Fig. 5.3). In chemistry and X-rays, we benefit from the skills of chemistry professors or researchers from Rennes, such as Jacques Lucas, Christiane and André Perrin, and members of the *Rennes en Sciences* association.

FIGURE 5.3 William Tobin and Alain Faisant reconstruct the experiment of the Foucault pendulum, Rennes

5.2 The "House for Science in Brittany at the Service of Teachers"

From the collections, we were able to design the project *Maison pour la science en Bretagne au service des professeurs* (House for Science in Brittany at the service of teachers).[21] Approved by the *Main à la Pâte* Foundation, supported by the Academy of Sciences and all of Brittany's teaching and research partners, it opened for the start of the 2014 school year, and is based on the scientific campus of Rennes, not far from the gallery of collections of scientific instruments.

At the start of each school year, the multidisciplinary training programme of the *Maison pour la science* is announced to all teachers in schools and colleges in the Brittany region. Courses of one to three days are offered every year to small groups of around ten teachers. The groups use the resources of the Rennes 1 University's collections of scientific instruments with visits to the gallery, the operation of devices and experiments, and replicas. For teachers in the primary cycle, who have little or no practice in scientific disciplines, practical and simple experiments offered during internships can boost their motivation or help overcome their fear of science.

Historical scientific instruments, by their simplicity, are excellent teaching aids that teachers can usually handle without too much difficulty or risk of deterioration. For example, to explore the concept of energy and its conservation, one can easily use a Lavoisier calorimeter, a Heron's fountain, study the operation of a James Watt steam engine, and attempt to perform very simple experiments by using these instruments as models.

The courses offered include:
– "The concept of Energy": conservation, transformation, transport.
– "Domesticated Energy": the history of its domestication, current challenges.
– "Let's get in tune!": sound at the crossroads of disciplines.
– "The progress of measurement": units, old and contemporary devices, precision, data acquisition and processing.

Teachers can manipulate some devices in the gallery and design small, simple experiments that they perform in a specific room at the *Maison pour la Science*. As these experiments require very simple and currently-available materials, they can easily reproduce them with their pupils when they return to their schools.

These exchange days are also an opportunity to make teachers aware of the issue of safeguarding and enhancing the scientific and technical heritage to which they may have access when they return to their colleges or high schools.

---

21 See: www.maisons-pour-la-science.org/fr/bretagne and https://maisons-pour-la-science .org/en (both accessed 5 Oct. 2020).

In June 2018, a large animation evening called *La Nuit des Profs* (Teachers' Night) was organised by the *Maison pour la Science* and the Cultural Service. It brought together about seventy teachers around animations and visits to the collections. One teacher's feedback shows how they experienced and appreciated this Teachers' Night: "A big thank you to the whole team for the beautiful evening spent with you. A 'Teacher's night' where I have felt pampered as much by the kindness of the reception as by the quality of the entertainment. There was joy in the air!"

6   Reception of School Children in the Collections

The reception of classes, from primary to high school, has always been considered an important activity of the cultural service because it puts young schoolchildren in contact with the university environment for the first time. We do little publicity because we have enough requests and our human resources are insufficient to meet the many enquiries. On arrival, the classes are divided into small groups from 10 to 15 students. By rotation, each group visits the different sections of the collection: scientific instruments, zoology and geology. The guided tour by a curator lasts between 40 minutes to one hour. The theme of each visit is tailored to the school programme, informed by an in-depth discussion with teachers prior to the visit.

It is widely illustrated by experiments sometimes carried out by pupils (when possible) and by demonstrations made by the curator. These meetings are an opportunity to make teachers and high school students aware of the need to conserve and promote contemporary scientific heritage and the importance of teaching the history of science (Fig. 5.4).

During these school visits we try to arouse astonishment and curiosity for these "very old instruments" in relation to the age of the visitors. We really like to show them devices that work without electricity, such as the slide rule, Archimedes' screw, the gramophone, Marey's recording cylinder or Heron's fountain (Fig. 5.5). These types of visit provide opportunities to discuss contemporary subjects with children, such as the evolution of technology, societal problems and energy issues. Some provocative questions are raised in these discussions: "Can we live without electricity? Do other energy sources exist? And how did humanity live before the Internet?" This interactive questioning makes it possible to tackle energy issues; the technological developments of inventions that are often associated with the waste of raw materials and pollution; or how to imagine life without the Internet.

FIGURE 5.4 Pupils and teachers in the instruments gallery; University of Rennes 1, 2014

FIGURE 5.5 When pupils are experimenting! Rennes, 2019

The collections department was able to accommodate up to 1,200 students in 2016 (24 classes). This maximum reception was achieved thanks to the implementation of a planned programme with the city of Rennes called *Accompagnement en Science et technologie à l'école Primaire* (ASTEP – Support for Science and Technology at Primary School level). This very substantial attendance record from 2016 of 1,200 students did not continue in the following years 2017, 2018 and 2019 (attendances of 500, 200 and 550 respectively) due to the drastic reduction of service personnel in the collections.

When high school students are presented with the evolution of computing methods, from the slide rule to desktop computers, including mechanical calculators, astonishment and surprise dominate. When we produce various forms of recording and storage of data from the cylinders of Marey (1900), 1930s music records to the USB sticks or the "cloud", the students realise how technology evolves in response to scientific discoveries.

7   Public and Scientific Instruments Heritage

The university, being a place essentially dedicated to the training of students and to research, is only open to the general public occasionally during events of dissemination of scientific culture. Among the main occasions, let us quote the *Fête de la Science* of Rennes (21,000 people including 2,000 schoolchildren over three days in 2019),[22] the European Night of Museums,[23] Heritage Days or the European Day of University Collections,[24] piloted by the UNIVERSEUM network (Fig. 5.6).[25]

The collections of instruments lend themselves well to all kinds of surprising or educational demonstrations. These festivals are the opportunity to present original experiments of historical and contemporary instruments. In this process, Paolo Brenni inspired us with the many experiments he has set up with his team at the Fondazione Scienza e Tecnica in Florence.[26]

---

22   See: www.espace-sciences.org/evenements/festival/2019/village (accessed 5 Oct. 2020).
23   See: www.univ-rennes1.fr/actualites/23042015/envie-de-passer-la-nuit-luniversite (accessed 5 Oct. 2020).
24   See: www.univ-rennes1.fr/node/1549 (accessed 5 Oct. 2020).
25   See: www.universeum-network.eu/ (accessed 5 Oct. 2020).
26   See: www.youtube.com/user/florencefst?feature=watch (accessed 5 Oct. 2020).

FIGURE 5.6 One poster of public demonstrations: Journée Européenne des Collections Universitaires; University of Rennes 1, 2014

THE COLLECTIONS OF SCIENTIFIC INSTRUMENTS IN RENNES 99

7.1   *The Reconstruction of the Historical Experiment of Pierre and Marie Curie*

Among the scientific activities presented to different audiences (students, schoolchildren and the *Maison pour la science*), the reconstruction of the experiment of radioactivity measurements by Pierre and Marie Curie figures prominently.[27]

With the help of Jacques Curie, they made wonderful discoveries and received Nobel Prizes: in 1903 for Pierre and Marie, and 1911 for Marie. The original electrical assembly combines an ionisation chamber containing the radioactive sample, quartz and an electrometer. This equipment is partly kept at the Ecole Supérieure de Chimie de Paris (ESPCI).

The devices designed by the Curies, such as quartz scale, were built and distributed by "La Société Centrale de Produits Chimique – SCPC" between 1885 and 1900. The physics professors at the Rennes science faculty acquired these devices when these first became available on the market between 1890 and 1900. In the 2000s, we thus found six historical devices in the Rennes collections: four quadrant electrometers, an electroscope, a quartz current generator (No. 2), a Curie-Blondlot wattmeter, an aperiodic balance, and a Curie-Cheneveau magnetic balance.[28]

The idea then came to us to try to reconstruct the original historical experiment with these devices. With the help of Bernard Pigelet, Denis Beaudouin and the staff of the Musée Curie, we were able to carry out this reconstruction, which was inaugurated in 2015 by Hélène Langevin and Pierre Joliot, grandchildren of Pierre and Marie Curie (Fig. 5.7).[29]

Installed in the geology gallery of the scientific campus, this experiment is part of the regular programme of visits. The various possible experiments are carried out directly by the curators: Julie Priser, Jean-Paul Taché and Dominique Bernard. The public can see the preparation (handling and positioning of very weak radioactive sources), the independent operation of the various devices,

---

27   See the video "L'expérience de Pierre et Marie Curie reconstituée à Rennes, novembre 2015" by Florence Riou, www.youtube.com/watch?v=Hsp5TljXzBU (accessed 5 Oct. 2020). This reconstruction obtained the second prize awarded by University Museums and Collections (UMAC) in 2017. See also: B. Pigelet and D. Bernard, "Reconstitution of the historical experience of measuring the radioactivity by Pierre and Marie Curie", *XXXVII Symposium of the Scientific Instrument Commission*, Leiden and Haarlem, The Netherlands, 3–7 September 2018.

28   The Rennes versions of the Curie devices are not radioactive and can be operated by museum people today.

29   See: "Voir la radioactivité: Reconstitution de la première expérience imaginée par Pierre Curie", www.youtube.com/watch?v=v0VHn3VbrZ8 (accessed 12 Mar. 2021).

FIGURE 5.7   The experiment of Pierre and Marie Curie reconstructed with quartz. The inauguration by B. Pigelet, H. Langevin and P Joliot; University of Rennes 1, 2015

and then observe on the light ruler the displacement of the spot proportional to the electric current.

The interpretive and experimental large public programme also presents books, documents, and videos. This general presentation deals also with the history of the great discoveries of the beginning of the twentieth century, the scientific method such as trial and error, nuclear energy and its applications. Curators, teachers and educators are always astonished to see the reactions of the public upon being shown historical devices in perfect working order. Even now, these artefacts exhibit remarkable sensitivities, such as electric currents measured at the order of a pico-ampere.

These programmes create the opportunity for very enriching exchanges with the public. Initially, the public is often frightened by the vision of physics and chemistry, considering these to be very difficult sciences. At same time, they are intrigued by the possibility of being able to understand better such important phenomena as radioactivity. Yet the reactions of those present are generally very enthusiastic, with the pleasure of seeing a live historical experiment and understanding how it works and its purpose. The educational interest of nineteenth-century scientific apparatus is highlighted once again.

FIGURE 5.8   Exhibition of different instruments from chemical laboratories at the Ecole Nationale Supérieure de Chimie de Rennes, 2019

7.2   *The Centenary of the National School of Chemistry of Rennes*

On the occasion of the centenary of its creation in 1919, the National School of Chemistry organised a whole series of public events, which ended with a big "open day". On this occasion Julie Priser, helped by the retired CNRS chemists André and Christiane Perrin, was able to set up, in a large practical working room, an exhibition of one hundred chemical analysis and measurement instruments that were commonly used in laboratories during the period 1950–2000. For example, we find an oil diffusion pump, a microbalance, a cryostat or X-ray apparatus. It is thanks to a collection started in research laboratories, within the framework of the PATSTEC program, that such a variety of objects could be kept, to the great pleasure of people present at this event (Fig. 5.8).

8   Research Laboratories and Conservation of Scientific Heritage

These examples of exhibition are quite typical of an awareness-raising effort undertaken by laboratories and researchers on the theme of the conservation of the scientific heritage. This subject is, of course, never a priority for those who, for their work, seek to equip themselves with the most efficient measuring devices and are often faced with problems of limited space and storage area available in their institutions.

However, we were able to maintain constant contact with a certain number of laboratories at the University of Rennes 1, in particular thanks to the network of associations of retired staff and with the support of the PATSTEC mission. This allowed us to perform "beautiful rescues".

The question of storing and moving large and very heavy instruments is not easy to resolve. It must be possible to find an exhibition site capable of receiving such devices in a place visible to students, staff and the public. This is an uncomfortable and rude question which is often settled by the rubbish bin, instead of by a thoughtful evaluation of the heritage and historical value of objects that have become obsolete for contemporary research.

Given the scientific and historical interest of certain instruments and with the support of administrative officials, we were able to present in the various halls of honour on the scientific campus some beautiful finds: two electron microscopes from a biology laboratory (a Philips EM 200-1967 and a JEOL 100 CX-1990); one of the first Varian mass spectrometers purchased in 1972 by the Western Regional Centre for Physical Measurements; and, more recently, in 2017, thanks to researchers from the Institute of Chemical Sciences (in particular Octavio Pena), we were able to save a Superconducting Quantum Interference Device (SQUID) magnetometer. Used to determine the magnetic properties of materials, this instrument was manufactured in very small numbers. The laboratory acquired this magnetometer in 1981.

Naturally, for each of these new acquisitions, we intend to compile a complete documentation of the technical and historical aspects, in particular by interviewing the personnel who used it routinely in the laboratory.

8.1 *A Particular Relationship between Scientific Instrument and Public*
Several specific situations are possible and interesting when we receive the public in the collection of scientific instruments of the University of Rennes 1. For example, we often encounter members of the public who arrive saying that they do not expect to understand anything: "since the visit is related to science, it is beyond me, too abstract for me". Another initial attitude is that of some students, who are reluctant to come, due to their assumption that the museum visit "will be long and tedious, and the objects are dusty or old-fashioned". A different form of disengagement is evident when a child, who is totally amazed by visiting the zoology gallery, upon arriving at the instruments, feels blunted and uninterested.

The question therefore arises: what is the relation between a museum visitor and the scientific object? This question, raised by the three concrete examples above, lays the very foundations of the interest of the use of the instruments of the nineteenth century in scientific pedagogy.

In France, since the nineteenth century, the object has been a support for science education. But over time, the relationship we have with instrumentation has changed completely. Modern apparatus, with its constitution and its presentation, adds a difficulty in the comprehension of a scientific idea. So, the associated scientific concept is put in a black box by the presenters of instructions related to the instrument, resulting in making both the concept and the instrument totally inaccessible.

By putting the nineteenth-century instruments back into use – maybe in a new or different way –, we make the scientific idea once again tangible. The historical object translated a conceptual subject in ways that still engage us directly today. Indeed, it surprises us and awakens our senses: we see an electric discharge, we hear (or not) a tuning fork, we participate in the show. The heritage object becomes educational because it allows us to perceive a scientific principle with the help of our senses and subsequently becomes a source of reflection. The observation of a phenomenon and its understanding are possible since the heritage object is constructed in such a way as to make a principle visible and evident to our human senses. The observer has the ability to formulate hypotheses and give initial answers to scientific questions such as: How does the object work? What is its use? The spectator becomes the actor by applying scientific reasoning. Added to this is the fact that the visitor is put at ease by the engaging way that the instrument animates the principles of science. Learning and understanding are all the more efficient.

Experimentation does not always give the expected or hoped-for results. These unforeseen outcomes therefore do not constitute a failure but provide new information that must be integrated in order to try to better understand the observed phenomenon. The advantage of nineteenth-century instruments is that the user is not helpless. Having understood easily and without prior knowledge of the operation of the object, which is often a mechanical object, she or he can suggest solutions to solve the problem. Here again, learning is valued. The public does not remain trapped by a failure, but is an actor in the process of ongoing investigation.

With the help of this instrumentation we can speak in the history of science and develop an attraction for science among members of the public. This experience of building a relation between the public and the instruments is generally a good introduction. It develops a certain curiosity and advances the popularisation of science.

Acknowledgements

All the activities described in this chapter are, of course, the result of collective work, which brings together the Brittany PATSEC mission, the Scientific Culture Commission, and the cultural and heritage service of the University of Rennes 1. We would like to particularly thank Jean-Paul Taché, Marie-Aude Lefeuvre, Marion Lemaire, André and Christiane Perrin, Jacques Lucas, Olivier Ridoux and Pierre-Antoine Angelini. Thanks also, for their collaboration, to student interns, laboratory researchers, the central documentation service, and colleagues from Maison Pour La Science or Rennes en Sciences Association.

We associate with these thanks all our friends and colleagues gathered in international networks like SIC, SIS, UMAC or UNIVERSEUM and their constant and unequalled help, which is invaluable to us.

CHAPTER 6

The Collection of Scientific Instruments from the Maraslean Teaching Center and Experimental Science Education: Then and Now

Panagiotis Lazos, Constantina Stefanidou and Constantine Skordoulis |
ORCID: 0000-0001-7517-656X; 0000-0002-2509-7764;
0000-0002-8748-1489

1   Introduction

The Greek state was founded in 1830.[1] The newly formed state lacked almost all the structures that characterise a state entity, and the establishment of primary schools, as well as their staffing with competent teachers, was imperative. In 1834, the Royal Teaching School (Βασιλικό Διδασκαλείο) was founded, in which teachers were trained. This school was succeeded by the Athens Teaching School (Διδασκαλείο εν Αθήναις) in 1876. In the new curriculum, the natural sciences occupied an important place, as was proved by the purchase of a significant quantity of high-quality instruments, and the appointment of a laboratory technical assistant.[2]

Indeed, the laboratory material book from the School's Laboratory (1899) mentions more than 300 instruments and devices for the experimental teaching of physics and chemistry, the oldest of which were purchased in 1879.[3] Among them were many prominent technological innovations of the time, and their existence highlights the progressive character of the school curriculum. The collection included, for example, an Edison phonograph, which was bought in 1890 – not long after its invention in 1877 – and two pairs of telephones from Bell and Siemens, as well as a Morse electric telegraph and other items.

---

1   This chapter originates from the presentation: Panagiotis Lazos, Constantine Skordoulis, "The Collection of the Scientific Instruments in Marasleios Pedagogical School: A Conflict between Past and Present", *XXXIV Symposium of the Scientific Instrument Commission*, Turin, Italy, 7–11 September 2015.
2   Konstantinos Tampakis, *Teaching Natural Sciences to Prospective Primary Teachers in the Greek State (1835–1950)*, Doctoral Thesis, National and Kapodistrian University of Athens, 2009, p. 91. See also: *Law 1040* [Νόμος ΑΜ'], June 29, 1882.
3   Βιβλίο Υλικού Εργαστηρίων Μαρασλείου [Maraslean Laboratory Material Book], 1899, MS.; in Tampakis, *Op. cit.* (n. 2), "Appendix", pp. 22–44.

© PANAGIOTIS LAZOS ET AL., 2022 | DOI:10.1163/9789004499676_008

2 The Maraslean Teaching Center and the Experimental Teaching of Natural Sciences

The Athens Teaching School was relocated in 1905 to a neoclassical building. The cost of constructing, furnishing and equipping this building was covered by the noted Greek national benefactor Grigorios Maraslis, from Odessa, Ukraine. As a result, in 1910, the school was renamed the Maraslean Teaching Center and it operated as such until 1933. At that time, the pedagogical academies undertook the training of teachers, and this included the Maraslean Pedagogical Academy, which was still housed in the neoclassical building funded by Maraslis (Fig. 6.1).

Dimitrios Gizelis, professor of physics at the Academy, reported in 1939 the instruments of the laboratory came from "all European physics instrument factories", but that most were now useless.[4] Even those that had been repaired and used for the teaching of students were not considered adequate (Fig. 6.2).[5]

In the school year 1937–1938, the Academy received from the Greek State the amount of 300,000 drachmas for the improvement of the laboratory. This amount was equal to about 170 workers' salaries, and corresponds to 200,000 euros today.[6] Part of this amount was spent on the purchase of 62 instruments and appliances manufactured by the German factory Phywe, fewer than twenty of which survive today.[7]

In 1985, students were admitted to the Maraslean Pedagogical Academy for its final year. Subsequently, after the pedagogical departments of the Universities were founded, these took over the training of primary education teachers. New physics laboratories were created in these pedagogical departments, and the laboratories of the Academy ceased to be used. The strong earthquake that hit Athens on September 7, 1999 left the building with serious structural problems, which led to it being evacuated and falling out of use. It is worth noting that the doctoral students at the time repeatedly entered the building to remove the instruments in the collection, as well as other valuable objects, risking

---

4 Yiorgos Palaiologos, *Ο θεσμός των παιδαγωγικών ακαδημιών και η Μαράσλειος Παιδαγωγική Ακαδημία* [The institution of pedagogical academies and the Maraslean Pedagogical Academy], Αρχαίος Εκδοτικός Οίκος Δημητρίου Δημητράκου Α.Ε. [Publisher Dimitrios Dimitrakos], Athens, 1939, pp. 321–337. It is noted that the equipment from the Corfu school had been transferred to the Athens school by 1899 at the latest, as it is mentioned in the laboratory material book of this school.

5 *Ibid.*, p. 330.

6 Stergios Babanasis, "Η διαμόρφωση της φτώχειας στην Ελλάδα του 20ου αιώνα (1900–1981) [The formation of poverty in Greece in the 20th century (1900–1981)]", *Επιθεώρηση Κοινωνικών Ερευνών* [*The Greek Review of Social Research*] 42 (1981), pp. 110–144: p. 111.

7 Βιβλίο Υλικού Εργαστηρίων Μαρασλείου [Maraslean Laboratory Material Book] 14/7/1937, MS; in Tampakis, *Op. cit.* (n. 2), "Appendix", pp. 22–27.

SCIENTIFIC INSTRUMENTS IN THE MARASLEAN TEACHING CENTER 107

FIGURE 6.1 The Maraslean Teaching Center around 1930
PHOTO BY PETROS POULIDIS ERT-ARCHIVES

FIGURE 6.2 Experiments with hydrostatic balances in the Maraslean Teaching Center, 1938–1939
YIORGOS PALAIOLOGOS, *THE INSTITUTION OF PEDAGOGICAL ACADEMIES AND THE MARASLEAN PEDAGOGICAL ACADEMY*, PUBLISHER DIMITRIOS DIMITRAKOS, ATHENS, 1939, P. 330

their lives in doing so. After extensive repairs, the building was reopened at the start of the school year 2009–2010, and the rescued instruments were placed in two rooms in the basement, where they remained out of use. This collection includes around 160 instruments and devices from European manufacturers such as Ducretet (Paris), Ernst Schotte (Berlin), Hartmann & Braun (Frankfurt), Leppin & Masche (Berlin) and E. Zimmermann (Leipzig – Berlin), but also several from the state-owned factory of school laboratory equipment in Athens, which closed in 1990.

The following paragraphs analyse the value of the preservation of historical scientific instruments of the Maraslean Teaching Center and the first attempts to introduce them to educational programmes.

2.1 *The Context*

In the school year 2014–2015, the first author of this chapter, in consultation with the third author, who was then president of the Maraslean Pedagogical Academy, began an effort to record, identify, clean and, where possible, repair the instruments. A group of students from the school where the former had taught the previous year expressed a desire to participate in this effort. The students had, for the two years prior to this, participated in a school activity programme entitled "School team for science experiments", and were particularly interested in scientific instruments.

Specifically, nine students (seven boys and two girls) from the second and third grades of the 26th General Lyceum of Athens (age 16–17) took part in the programme. Their participation was voluntary, non-graded and informal, in the sense that it did not take place within a school programme or during school hours. The meetings took place in the newly established Laboratory of Natural Sciences in the building of the Maraslean Academy. The students went there immediately after their lessons ended.[8]

The original idea of the initiative was exclusively centred on cleaning, identifying and arranging the instruments, with students providing valuable assistance but not setting any educational goals. In other words, the main focus was on the instruments themselves. However, very quickly and even before the initiative had started, the focus shifted to some degree from the instruments onto the students.

Following a proposal by Efthymios Nikolaidis, who was teaching "History and Philosophy of Science" – a course from the autumn semester of the Pedagogical Department's postgraduate programme "Natural Sciences in Education" – it

---

8 There was, of course, prior consultation with parents and guardians regarding the students' participation in the programme.

SCIENTIFIC INSTRUMENTS IN THE MARASLEAN TEACHING CENTER

was decided that seven second-year postgraduate students (six females and one male) would participate in the programme. These students were graduates of the Pedagogical Department and possessed both a good level of knowledge and a specific interest in the natural sciences. Their participation was accompanied by the preparation of weekly assignments, the content of which is analysed below, and a final assignment in which each group described in detail the experiments and scientific instruments that appeared in Greek textbooks for experimental physics at the end of the nineteenth century.

A total of fifteen meetings were held between November 2014 and March 2015. The high school students participated in all the meetings, while the postgraduate students took part in six of them (meetings four to nine) corresponding to half a semester, as in the remaining semester they attended lectures on different topics.

2.2 *The Structure of the Educational Programme*

Activities with the students: For each meeting, several instruments were chosen beforehand, and these were assigned to groups of two or three students so that, by the end of the final meeting, nearly all of the instruments in the collection had come under "study".

We used an approach that shows similarities to the Winterthur model;[9] however, it was developed independently. The groups filled out a worksheet, which asked them:

– to describe the instrument as fully as possible (dimensions, materials, structure; Fig. 6.3);
– to describe any deficiencies or damage they could identify;
– to find out any information on the instrument about its manufacturer and date of manufacture and, if not possible, to suggest alternative ways of finding such information;
– to suggest the possible intended use of the instrument and to decide which "branch" (or branches) of physics it would have belonged to, judging from its structure and characteristics;
– to indicate whether they believed the instrument worked autonomously or in combination with others in an experimental setup and, in the case of the latter, with which other instruments;
– to try to suggest an experiment involving the specific instrument and come up with its corresponding scientific question.

---

9 E. McClung Fleming, "Artifact Study: A Proposed Model", *Winterthur Portfolio* 9 (1974), pp. 153–173.

FIGURE 6.3 A team of students study a Müller apparatus for the laws of reflection
PHOTO BY P. LAZOS, 2014

At the end of each worksheet, which was designed by the instructor, information was given about the instrument, usually taken from trade catalogues or physics textbooks of this era. The students were thus able to check their initial predictions and answer questions that arose while they were exploring the instrument and discussing it. In the case of instruments in good working condition the students used the instrument under the guidance of the instructor.

For example, they used the stereoscope to see old stereoscopic photographs.[10] Their surprise was great as they realised not only that the theory worked, but the result was impressive, providing a three-dimensional picture in the middle of the nineteenth century.

There was also an effort to find alternative ways to illustrate how an instrument worked if the instrument itself was not usable, as it was with the Davy lamp. The students used a candle and a wire mesh in order to see and understand that the flame stops under the mesh.

Where possible, additional instructions were given for the students to be able to construct a similar instrument using simple materials. They were not, however, asked to construct it because this was out of the scope of the project.

In addition, students were asked to suggest ways of repairing the instruments where possible. The condition for this was that the repairs should be easily reversible and non-intrusive, in order to ensure the authenticity of the instrument while at the same time making it useable. The students suggested and carried out two such repairs. The first involved a Wimshurst machine built by the state-owned factory of school laboratory equipment in Athens, which operated from 1950 to 1990.[11] New straps were inserted into the device and the broken Leyden jar was replaced with another made from a plastic cup, tin foil sheets and a chain (Fig. 6.4).

The students already knew, from similar repairs in the school laboratory during the previous school year, how to use a specific strap for the repair of a Wimshurst machine. They did an internet search for the Leyden jar and then they chose the necessary materials. The machine became fully usable – although probably not as beautiful – and can be returned to its original state within a short time.

The second repair was made to a device that initially had three pendulums, all of which were missing. The students studied the description and design of similar devices in trade catalogues, and found that the lengths of the three pendulums had a ratio of 1:4:9, so that the respective periods had a ratio of 1:2:3. The students were then left to decide what materials to use to repair the device. After some tests (for example, with fishing weights or nuts), they

---

10   Unfortunately, there are no surviving stereoscopic photographs in the Maraslean collection, so the students used photographs provided by the instructor.

11   For information on this see: Kostas Kambouris, Κέντρο Εποπτικών Μέσων Διδασκαλίας (1950–1990). Η συνεισφορά του στην Πειραματική διδασκαλία των Φυσικών Επιστημών στην Εκπαίδευση [The Center for Teaching Equipment (1950–1990). Its contribution to the Experimental teaching of the Natural Sciences in Education], [K. Kambouris], Athens, 2006; available on line at: https://panekfe.gr/downloads/ekfe/2006-kampouris-to-kentro-epoptikwn-meswn-didaskalias-1950-1990.pdf (accessed 21 Jan. 2021).

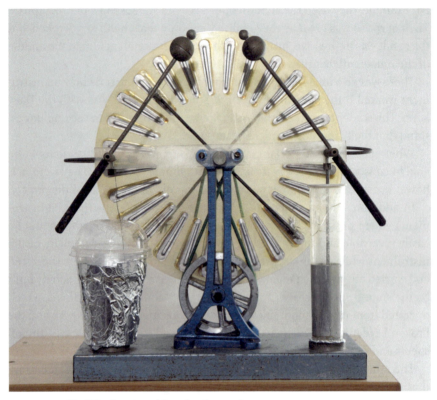

FIGURE 6.4 The Wimshurst machine after the repair
         PHOTO BY P. LAZOS, 2020

decided upon using thin thread and metal beads from a Newton pendulum (Fig. 6.5). The activity was completed using the layout and confirming the ratio of periods. This device can also be returned to its original state very easily.

Activities with postgraduate students: The activities with the postgraduate students were, in general, similar to those carried out with the high school students (group work, worksheets on specific instruments, similar approach and questions, and so on). However, the smaller number of meetings necessitated a more rigorous selection of instruments. Because these students had chosen to pursue professions in education, instruments and activities were selected that highlighted the relationship between instruments and the scientific, technological and cultural situation of their time. The purpose was for the students to, on the one hand, experience that science is not an isolated field, while on the other hand, develop a familiarity with and a positive attitude towards the material culture of science.

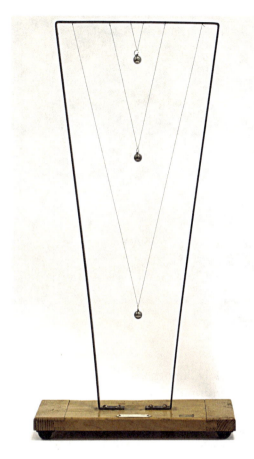

FIGURE 6.5   The three pendulums after the repair
            PHOTO BY P. LAZOS, 2015

In addition to the activities that have already been described, the students were encouraged – in the context of their weekly work on specific instruments – to explore somewhat more unusual aspects of the instrument, such as:
– to explain why so few records from phonographs have been saved;
– to present the similarities and differences between a modern AA battery and a Grenet or Leclanché electric cell from the nineteenth century;
– to find advertisements for Davy lamps in magazines and newspapers of the time;
– to locate used stereoscopes on eBay and compare their structure with the structure of the stereoscope of the Maraslean collection;
– to describe in which technological applications the electric arc was used.

FIGURE 6.6   The instruments in the laboratory
PHOTO BY P. LAZOS, 2020

This part of the research was done mainly through the internet and in some cases it led the students back to the instruments in order to examine or to verify something they had found. That was the case for the electric arc. The students wanted to examine whether there was a manual or an automatic system for maintaining the correct distance between the two electrodes (a piece of information they had not found through the worksheet). The Davy lamp was another example as the students, after finding various images of Davy lamps in advertisements, decided to search for differences between them and the sample of the Maraslean collection. This way of "back and forth" study was not planned but it came about spontaneously as the process progressed.

At the end of the project all the instruments of the collection were recorded, identified, cleaned and placed in and on display cases in the laboratory (Fig. 6.6).

2.3   *Evaluation of the Educational Programme*
No rigorous data collection and processing methodology was used during the programme, at least where the students were concerned. Instead, the programme followed the characteristics of a research initiative in which the teacher participated in all phases of the project in order to coordinate groups,

observe action, pose questions, explain difficult technical features (being aware of the level of both groups of students) and, of course, protect the instruments from possible damage.

The data collection tools were the students' worksheets, work produced by the postgraduate students, a diary which was kept, and discussions which took place during the meetings. Finally, open-ended questionnaires were given to all participants five years after the implementation of the educational programme. An analysis of students' worksheets and the teacher's diary show that, over the course of the programme, students from both groups:

– were able to understand what the necessary aspects of the instruments are for a complete description (for example: dimensions, material, inscriptions, and so on) and to give accurate information about them;
– faced some – rather anticipated – difficulties in giving approximately accurate predictions about the three last questions in the worksheets (a prediction about the use of the instrument, its autonomous function or not, and a formulation of a hypothesis for experiments in which the specific instrument could participate). Sometimes, especially in the early meetings, participants did not even attempt the questions. Nevertheless, there were some exceptions, usually when the students had had a previous experience with a relevant instrument;
– gradually began to distinguish – albeit approximately – the date of manufacture of the instruments judging from their external features and materials;
– could suggest which of the instruments they had already studied could be combined with instruments they were studying at a given time;
– were particularly impressed by some of the instruments. One example is the Davy lamp and how its clever design saved the lives of many miners. Moreover, the simple idea behind the acoustic siren and the fact that the civil defence sirens in Athens work based on the same principle was also quite striking (Fig. 6.7);
– gained the ability to draw useful information on the scientific instruments by studying the commercial trade catalogues from manufacturers and locate objects that are held in our collection. In one case, a group of students managed to find in the Max Kohl catalogue of 1905, a depiction of the four-flame gas burner designed by Finkener; this instrument in our collection had not previously been identified with such accuracy by the authors of this paper;
– developed feelings of respect and admiration for both the instruments and the people who designed, manufactured and used them;
– perceived that scientific instruments are not only associated with science and technology but also with other social activities, such as trade and art.

FIGURE 6.7 A team of students trying to conceptualise how the acoustic siren works
PHOTO BY P. LAZOS, 2014

    Through research on the makers of scientific instruments, the students discovered an extensive trade network with international exhibitions, trade representatives in many countries and catalogue publishing;
- realised that several of the modern technological innovations about art have their roots in objects that they could touch and study. The phonograph, the birth certificate of recording and reproduction of sound, and the stereoscope, which incorporates the principle of the operation of 3D cinema, were tangible before their very eyes.

In terms of co-operation and collaboration between students and postgraduate students, as all graduate student meetings took place alongside those of the high school students, there was often interaction between groups, sometimes planned and sometimes spontaneous. This experience was excellent for the younger students, who were able to glimpse their own future as postgraduate students, while also learning from their methodical and careful ways of working. In turn, postgraduate students had the opportunity to become teachers in an informal setting, and were often pleasantly surprised by the high school students' ideas and level of knowledge.

    In an interesting incident, a team of high school students was studying a long glass tube for the mercury fine shower demonstration. The students immediately recognised the object, as they had seen a similar one during an educational

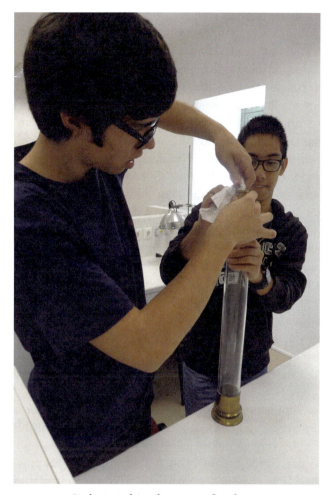

FIGURE 6.8   Students studying the mercury fine shower apparatus
PHOTO BY P. LAZOS, 2014

visit to the Museum of Geoastrophysics of the National Observatory of Athens. The postgraduate students were quite impressed by this. Nevertheless, they were even more impressed when the high school students explained that a piece of chalk and a piece of paper inside the tube indicated that the tube was used for free fall demonstration in a subsequent demonstration (Fig. 6.8).

In order to have an idea about the impact that the educational programme had on the participants, we sent by email five questions to all of them during July 2020, five years after the implementation of the project. Eight out of nine former high school students and only two out of seven postgraduate students answered. Seven out of eight students have followed science studies and the two postgraduates are primary teachers with a master's degree in science education. The questions asked of both groups were related to: 1. positive thoughts about the programme; 2. what they liked most in the procedure; 3. their views on how the programme helped them with the maintenance of historical scientific instruments; 4. the programme's positive impact on their studies or daily life; and finally, 5. if they would participate in a future similar programme.

Regarding the first question, about their clearest thoughts regarding the project, all respondents expressed highly positive memories which had to do with the excellent atmosphere of the building and the particular room in which the historical scientific instruments were stored, besides the excellent atmosphere in the team and their relationship and collaboration. Indicatively, with a questionnaire participant's words: "I will never forget the room in which the instruments were stored [...] the atmosphere [...] it was a room full of treasures".

Regarding the second question, which had to do with students' favourite phases during the procedure, all individuals referred to the phase of maintenance and the phase that they managed making the instrument work again. They enjoyed the investigation of instruments' use and function. In a participant's words: "As a humanity orientation student I couldn't grasp all details about how instruments worked, but I loved seeing them re-live after our careful repair". Another respondent pointed out: "I really enjoyed the feeling of exploration as we did not know from the beginning what the instruments were made for".

In the context of the third question, individuals had to recall phases where they felt that what they were doing in the laboratory was useful not only for themselves but also for the very continuation of the existence of these scientific instruments. Regarding the personal benefit, most respondents mentioned that they realised how physics is everywhere; since all principles of function are present everywhere. Moreover, they gave great emphasis on the fact they feel the satisfaction of giving life to an old instrument. In a questionnaire participant's words: "I will never forget the feeling when you see an instrument before and after; an instrument locked for decades in a dark corner. It was extremely enjoyable to bring to life a scientific instrument that no one had used for decades". Another participant expressed the opinion that the programme had a positive impact on continuing the preservation of scientific instruments, since the work done could be a base for future research. Such an opinion shows that students respected the results of their research, as it is revealed through other words: "It is very satisfactory to 'clean dust' and 'repair' the past in order to have the future".

Regarding former students' opinion about how the programme influenced their daily life and/or studies, some of them answered that this influence is direct, while for some of them, it is indirect. For example, a participant said: "In the context of my studies, I have a lot of work in the laboratories. Here I can find several instruments, some of them new, some of them old. One of the lessons learned from the project is that old instruments may be improved if they are preserved and repaired properly. This makes their life longer [...] This is something I will never forget after the Maraslean project".

Finally, all individuals were willing to participate in a similar future programme mainly because they had a really good time, combining fun, researching and learning. All of them recognised that the Maraslean project helped them realise that science education had a position in primary teachers' curricula at least since the end of the nineteenth century. They found it motivating for their future careers, as most of them followed a science career. Besides other arguments for their future participation in a similar project, they all stated that they had the opportunity to express their creativity during the project, while the project combined entertainment and learning.

Postgraduate students' responses were consistent with those of the former high school students. They answered two more questions regarding their professional development. The first additional question had to do with the impact the programme had on their professional development in general. Both postgraduate students are primary school teachers and answered that the programme helped them to surpass their difficulties and apprehension about experimental physics. The last question was about the extent to which the programme supported their everyday teaching practice by using the school laboratories. Both responders attributed to the programme their increased self-confidence and their developed ability to manage with scientific instruments. One of them was so enthusiastic as to write: "The Maraslean project helped me a lot in organising and utilising the laboratories of the schools in which I have worked. I am more familiar with scientific instruments. I managed to repair and organise the scientific instruments of the abandoned laboratory at the 39th Primary School of Athens, where I worked in the school year 2014–2015, with your valuable help! My interest in this process was genuine, sincere and was provoked through the workshops you did for us! Your help was invaluable and the time you devoted was very important. Thank you again for that, no matter how many years have passed!"

These findings are particularly encouraging, especially considering that our main goals were to engage, observe, describe, collaborate and gradually familiarise participants with the world of scientific instruments.

3 Future Perspectives

So far, educational applications of the historical scientific instruments of the Maraslean Teaching Center have been presented to high school and postgraduate students. Our experiences from these training programmes feed into the planning of further educational programmes for the future. Our plans for teacher education take into account current trends for an interdisciplinary approach towards science, technology, engineering and mathematics (STEM), the need for teacher training in STEM, as well as the need to raise awareness of issues related to the conservation and utilisation of our scientific heritage.

In this context, we are planning the design and implementation of an in-service science teachers' STEM education programme. This will be based on the preservation and reconstruction of scientific instruments, followed by the development of educational activities for both students and the public, as the project will include a digital display of the collection of historical scientific instruments for further accessibility. The project under consideration is entitled "Historical Scientific Instruments in the Context of STEM Education (Hi-Sci-STEM)" and has been submitted as a proposal for funding to the Hellenic Foundation for Research and Innovation.

4 Conclusions

This paper mainly covered the following objectives: 1. to briefly present the history of the Maraslean Teaching Center focusing on the collection of scientific instruments and its preservation, and 2. to describe vividly the first attempts to use this collection in the context of science education.

This was the first attempt for educational use, and it was implemented with secondary students (16 to 17 years-old) and postgraduate students. Both teams were engaged in the recording, identification, preservation and repair of the instruments, their history, their function, and their use in science class. The findings reveal that, despite the conceptual difficulties students may have regarding the principles of function of some of the instruments, they were willing to participate in the process, as they found the process pleasant and useful not only for themselves, but for the well-being of the instruments as well. Regarding the former high school students' participation, they were enthusiastic making the instruments work and enabling future historical research. In their own words, the programme made them more creative. Regarding the postgraduate students, they found the Maraslean project helpful for the preservation of historical scientific instruments and the research of local history of science education. Moreover, postgraduate students stated they were more prepared to manage in a modern school laboratory. All students said they were willing to participate again in a similar project.

Therefore, the most crucial contribution of this paper is that although findings are limited to the sample, they provided strong evidence that local collections of historical scientific instruments, such as the Maraslean collection, may be an appropriate context for students' familiarisation of scientific and technological culture.

CHAPTER 7

Examples of the Use in Education of Historical Physics Instruments at Secondary School and University Level in France Supported by ASEISTE

Françoise Khantine-Langlois, Alfonso San-Miguel and Pierre Lauginie |
ORCID: 0000-0001-8013-2246; 0000-0002-3147-4116;
0000-0003-1112-4511

1 Introduction: Context of the Actions

We present here the collaboration between the *Association de Sauvegarde et d'Étude des Instruments Scientifiques et Techniques de l'Enseignement* (Association for Preserving and Studying the Scientific and Technical Instruments of Education – ASEISTE) and professors of physics in schools and universities. This collaboration develops the use of historical physics instruments held at these schools with pedagogy projects that involve students at the schools. The ASEISTE is a French association, founded in 2004, aiming at the preservation of the scientific and technical instruments of education. Six universities and sixty secondary schools, all having their own collections of scientific instruments, are involved in ASEISTE activities, either as associated partners or as institutional members. The main objectives of ASEISTE are: to rescue and preserve instruments and collections; to create comprehensive catalogues, both on-line or in printed books; and to develop pedagogical projects involving this scientific historical heritage in collaboration among members and professors.

The following section briefly presents the ASEISTE association. The body of this chapter develops three examples of pedagogical projects involving historical physics instruments: two at secondary school level and another involving Master's students at the university.

© FRANÇOISE KHANTINE-LANGLOIS ET AL., 2022 | DOI:10.1163/9789004499676_009

2 Presentation and Evolution of ASEISTE

In a previous article, we presented the ASEISTE as well as its website.[1] Two major projects successfully completed since then: the expansion of the association website and the edition of an encyclopaedia of physical instruments.

The ASEISTE website includes a comprehensive catalogue of historical instruments used in physics and chemistry. Presently, the number of instrument files has grown to 7,500. An important video collection of different sources is also referenced, including presentations from speakers such as Paolo Brenni, Bruno Jacomy, Marta Lourenço, Sofia Talas, Małgorzata Taborska or Yves Quéré (member of the French Academy of Sciences) and many others. Consultation of the site is increasingly growing, with 5,275 visitors corresponding to 8,286 visits and 44,390 pages viewed since 2019.

The *Encyclopédie des instruments de l'enseignement de la Physique du XVIIIe au milieu du XXe siècle* (Encyclopaedia of physics instruments developed for pedagogical purposes from the eighteenth to the twentieth centuries) has been published by the ASEISTE in December 2016. The main body of the encyclopaedia comprises a selection of almost 1,000 typological cards of instruments. Fig. 7.1 shows an example of a typological card corresponding to an apparatus for the study of cycloids. In addition, the encyclopaedia includes a collection of more than 200 postcards from the beginning of the twentieth century showing physics cabinets and instruments or demonstrations during lectures. Fig. 7.2 is an example of one of the postcards illustrating a demonstration. The three-volume work was written under the direction of Francis Gires, founder and past president of the ASEISTE.[2]

In addition, we mention the use of one of the objects of the ASEISTE collections – the Drummond lamp, or "limelight" – in a scene of the film *Les magiciens de la lumière* (Wizards of Light, 2009), which relates the measurements of the speed of light from Galileo to Léon Foucault.[3] The Drummond lamp was Hippolyte Fizeau's light source in his famous 1849 measurement

---

1 Francis Gires, Pierre Lauginie, "Preserving the Scientific and Technical Heritage of Education: The ASEISTE (www.aseiste.org)", *Bulletin of the Scientific Instrument Society* 121 (2014), pp. 35–41.
2 Francis Gires (ed.), *Encyclopédie des instruments de l'enseignement de la physique du XVIIIe au milieu du XXe siècle*, ASEISTE, Niort, 2016.
3 Pierre Lauginie and Alain Sarfati, *Les magiciens de la lumière* (DVD with French and English subtitles), SCAVO, Université Paris-Sud, 2009 (also available at: www.canal-u.tv/video/scavo/les_magiciens_de_la_lumiere.18096; accessed 30 Oct. 2020).

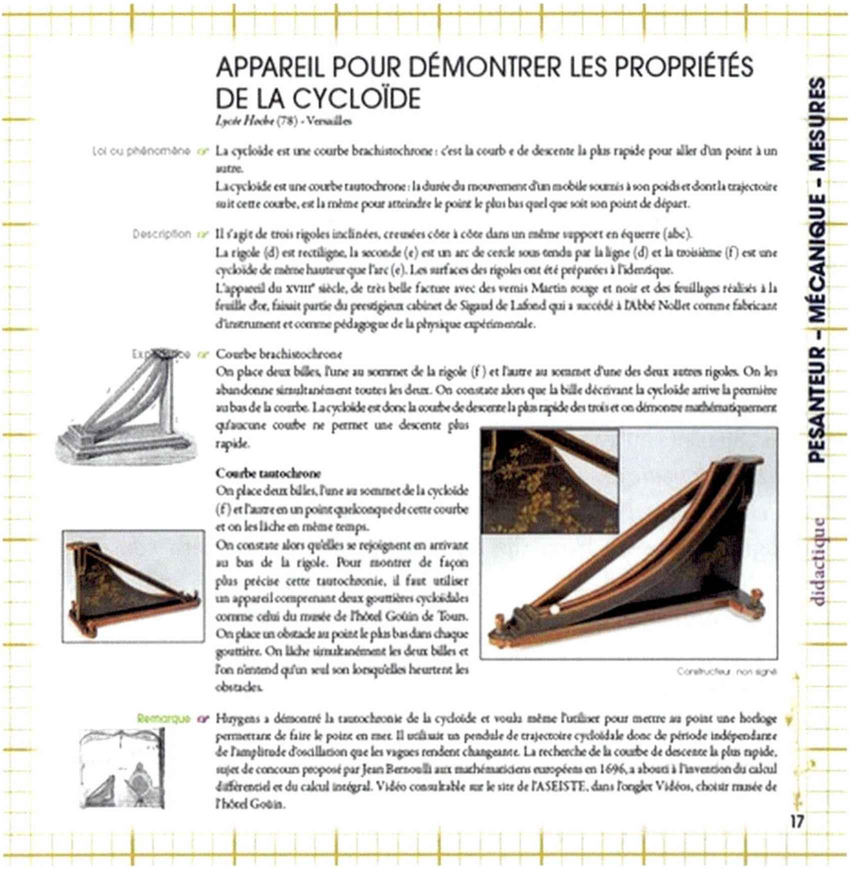

APPAREIL POUR DÉMONTRER LES PROPRIÉTÉS DE LA CYCLOÏDE

Lycée Hoche (78) - Versailles

Loi ou phénomène La cycloïde est une courbe brachistochrone : c'est la courb e de descente la plus rapide pour aller d'un point à un autre.

La cycloïde est une courbe tautochrone : la durée du mouvement d'un mobile soumis à son poids et dont la trajectoire suit cette courbe, est la même pour atteindre le point le plus bas quel que soit son point de départ.

Description Il s'agit de trois rigoles inclinées, creusées côte à côte dans un même support en équerre (abc).

La rigole (d) est rectiligne, la seconde (e) est un arc de cercle sous tendu par la ligne (d) et la troisième (f) est une cycloïde de même hauteur que l'arc (e). Les surfaces des rigoles ont été préparées à l'identique.

L'appareil du XVIIIe siècle, de très belle facture avec des vernis Martin rouge et noir et des feuillages réalisés à la feuille d'or, faisait partie du prestigieux cabinet de Sigaud de Lafond qui a succédé à l'Abbé Nollet comme fabricant d'instrument et comme pédagogue de la physique expérimentale.

Expérience Courbe brachistochrone

On place deux billes, l'une au sommet de la rigole (f) et l'autre au sommet d'une des deux autres rigoles. On les abandonne simultanément toutes les deux. On constate alors que la bille décrivant la cycloïde arrive la première au bas de la courbe. La cycloïde est donc la courbe de descente la plus rapide des trois et on démontre mathématiquement qu'aucune courbe ne permet une descente plus rapide.

Courbe tautochrone

On place deux billes, l'une au sommet de la cycloïde (f) et l'autre en un point quelconque de cette courbe et on les lâche en même temps.

On constate alors qu'elles se rejoignent en arrivant au bas de la rigole. Pour montrer de façon plus précise cette tautochronie, il faut utiliser un appareil comprenant deux gouttières cycloïdales comme celui du musée de l'hôtel Goüin de Tours. On place un obstacle au point le plus bas dans chaque gouttière. On lâche simultanément les deux billes et l'on n'entend qu'un seul son lorsqu'elles heurtent les obstacles.

Constructeur non signé

Remarque Huygens a démontré la tautochronie de la cycloïde et voulu même l'utiliser pour mettre au point une horloge permettant de faire le point en mer. Il utilisait un pendule de trajectoire cycloïdale donc de période indépendante de l'amplitude d'oscillation que les vagues rendent changeante. La recherche de la courbe de descente la plus rapide, sujet de concours proposé par Jean Bernoulli aux mathématiciens européens en 1696, a abouti à l'invention du calcul différentiel et du calcul intégral. Vidéo consultable sur le site de l'ASEISTE, dans l'onglet Vidéos, choisir musée de l'hôtel Goüin.

PESANTEUR - MÉCANIQUE - MESURES

didactique

17

FIGURE 7.1 Example of a file of the *Encyclopédie des instruments de l'enseignement de la Physique du XVIII*e *au milieu du XX*e *siècle* (Encyclopaedia of Physics Teaching Instruments from the eighteenth to the twentieth centuries): instrument to demonstrate the properties of the cycloid

with the toothed wheel, and it is also behind Charlie Chaplin's film *Limelight* (1952). The Drummond lamp was closely related to scientific, technological, artistic and political aspects of nineteenth-century society.[4]

In addition to these activities, the members of the ASEISTE use this heritage for educational purposes. In the following sections, we present examples of this activity, both in secondary schools and at universities.

4 Pierre Lauginie, "Drummond Light, Limelight: A Device of its Time", *Bulletin of the Scientific Instruments Society* 127 (2015), pp. 22–28.

FIGURE 7.2 Demonstration with physics instruments at the Lycée Hoche of Versailles, postcard, beginning of the twentieth century
ENCYCLOPÉDIE DES INSTRUMENTS DE L'ENSEIGNEMENT DE LA PHYSIQUE DU XVIIIE AU MILIEU DU XXE SIÈCLE, ASEISTE, 2016

3 Projects with Secondary School Students

History of science is included in today's French secondary school curricula. Students, from 11 to 18 years old, are asked to read short historical texts from, for example, Ampère or Lavoisier. In fact, the education curriculum for middle school pupils (11–14 years old) requests that scientific and technological developments be placed in a historical, geographical, economic and cultural context. In particular, it is explicitly stated that "it is possible to propose projects based on scientific or navigation instruments".[5] Moreover, for final year students in secondary school (17–18 years old), physics professors have at their disposal guides for the setting up of historical devices illustrating the properties of fluids based on the experiences or historical contributions of Bernoulli, Euler or d'Alembert. However, and despite this official recommendation, the experimental aspect is seldom taken into account. This is the reason that motivated ASEISTE to offer its support to teachers who want to exploit the heritage present in their own schools. Obviously the first step is to identify and preserve

---

5 BO-Bulletin Officiel de l'Éducation Nationale n. spéciale 11 (26 Nov. 2015), p. 338.

the school collections. For that purpose, the ASEISTE website is a very useful tool. By putting teachers in direct contact with members of the association, it helps them to: identify valuable objects, take high quality photographs, and write descriptive cards. In addition, the ASEISTE members may provide advice tailored to preserving particular scientific instruments. For very valuable or unique ones, the ASEISTE organises the protection of the instruments through a process involving the French Ministry of Culture.

An average estimation of the size of the collection of instruments in each of the sixty secondary schools associated with ASEISTE can be obtained by considering the number of filled files which have been included in the data-base platform. The average number is 65 instrument files per secondary school, with a median of 46 files. Remarkably, sixteen secondary schools have filled out more than one hundred files each, with one school reaching 337 files on its own. These strikingly high numbers are associated with secondary schools created in the Napoleonic period for which the teaching programmes of the time included the need to acquire instruments based on an official list.[6] Most of these schools have some instruments exhibited in showcases. A few schools, having larger collections, dedicate a room for exhibitions, sometimes having a status close to a local museum.

In the following text, we present two different uses of historical instruments in two different schools: the Collège du Chateau of Morlaix (11 to 15 year old children) and the Lycée Hoche of Versailles (11 to 20 year old students). We have deliberately chosen two schools with very different social and geographical environments.

The Collège du Château of Morlaix, a small town of 15,000 inhabitants in Brittany, is a secondary school which possesses a small exhibition room for historical scientific instruments.[7] Many of the instruments are used by the professors during lectures to illustrate physical phenomena or physical laws: historical ammeters and voltmeters, in wood or Bakelite, are used to verify Ohm's law; the experience of Oersted is reproduced using an historical set-up; or an ancient Franklin hand boiler illustrates thermodynamic principles. The students appreciate different aspects of these demonstrations. They may be surprised to realise the degree of precision in measurements done with historical apparatus when compared with modern ones. Others realise with amazement that physics was first developed without the need of digital technology, or are

---

6   Bruno Belhoste (ed.), *Les sciences dans l'enseignement secondaire français – Textes officiels*, INRP, Paris, 1995: v. 1 – "1789–1914", pp. 98–102.

7   Some views of the College collection as well as its link with the ASEISTE can be found at: http://collegeduchateaumorlaix.fr/crbst_144.html (accessed 24 Apr. 2020).

FIGURE 7.3 Baroscope from the secondary school Collège du Château at Morlaix, France
PHOTO JEAN-YVES BLAISE

just intrigued by the existence of this heritage at their school. Some children may be just charmed by the beauty of the instruments. The collection can also be visited by an external public during open doors events. In that case, the students themselves take the role of guides. This includes European Heritage Days (EHD) where all visitors are welcomed, as well as parents, and the future students' "discovery days" where parents and future students visit the school. In addition to the guided tours, the physics teacher of the college of Morlaix may take the opportunity to make a small demonstration or explanation based on a particular instrument. As an example, a baroscope from the end of the nineteenth century (Fig. 7.3), held in the college collection, is used to explain Archimedes' principle to 10- and 11-year-old pupils who have come to visit their future school. The school includes groups of students with cognitive handicaps who follow a technical or professional curriculum, particularly in carpentry. These students have contributed to preserving the technical and scientific heritage of their school by constructing various wooden showcases.

Our second example, the Lycée Hoche of Versailles,[8] possesses a remarkable collection of physical instruments. The ASEISTE website includes 281 objects from this collection. The association *Les amis du musée historique du Lycée Hoche* (Friends of the historical museum of the Hoche high school) was created in 2009 to participate in the management and development of the historical teaching heritage of the Lycée Hoche, which includes an important collection of physics measurement apparatus.[9] The Lycée welcomes middle and high school students, as well as preparatory classes for engineering schools (corresponding to the first two years of university level), aged from 11 to 20 years old. All levels and all subjects exploit the scientific collections of the Lycée, illustrating the richness that a museum of scientific instruments represents for a school, and the many educational uses that can be made with it.

The school dates from the nineteenth century, but its collection of physical instruments is enriched by inherited instruments from previous noteworthy collections.[10] Of particular significance are four devices built by Jean-Antoine Nollet (1700–1770), or produced according to his principles, which originate in *Le cabinet de physique des enfants de France* (The children of France's physics cabinet), a collection of scientific instruments devoted to the instruction of the children of the French royal family. The four devices were ordered by *le Dauphin*, son of Louis XV and father of three kings (Louis XVI, Louis XVIII and Charles X) for the education of his sons. Other instruments now held in the collection of the Lycée Hoche came from the splendid collections of Marquis Armand-Louis de Serent, a rich politician and collector of the eighteenth and nineteenth centuries. The remaining instruments making up the Lycée Hoche collection were purchased by the school itself to respond to the nineteenth-century curricula. One can appreciate the richness of the Lycée Hoche collection through viewing the two nineteenth-century postcards reproduced in Figs. 7.2 and 7.4.

---

8   All the following information was provided by Christine Dalloubeix, physics teacher in PCSI (Physics, chemistry, engineering science curriculum) at the high school, member of the ASEISTE, and president of the association *Les amis du musée historique du lycée Hoche*.

9   Details on the history and the objects of the collection can be found at: www.amismusee hoche.fr/histoire-lycee/histoire-lycee.php (accessed 24 Apr. 2020).

10   See: www.amismuseehoche.fr/collection-sciences-physiques/origines-collection -sciences-physiques/origines-collection-sciences-physiques.php (accessed 24 Oct. 2020).

FIGURE 7.4   Physics cabinet at the Lycée Hoche of Versailles, postcard, beginning of the twentieth century
ENCYCLOPÉDIE DES INSTRUMENTS DE L'ENSEIGNEMENT DE LA PHYSIQUE DU XVIIIE AU MILIEU DU XXE SIÈCLE, ASEISTE, 2016

The French physicist Jules Antoine Lissajous (1822–1880), famous for his work on vibrations,[11] studied in this school. His memory is honoured in the physics lectures in this school. When making reference to the method invented by this physicist for measuring the frequencies, the teacher asks the students to *passer en Lissajous* (change to Lissajous), instead of making the usual request of using the X–Y function of the oscilloscope.

Owning an important collection of historical instruments, as is the case of the Lycée Hoche, constitutes a favourable context providing many opportunities to develop projects with the students which are widely exploited by teachers:

– Students write articles in the *Journal des collégiens* on the objects chosen from the Lycée museum and displayed in three of the high school's libraries. For this, they receive support from the Lycée librarian.
– Several times a year, the museum receives groups in the final year of elementary school from other institutions. At the end of the project, the elementary school pupils guide their parents through the museum on a visit. This action, made in collaboration with the Versailles city administration,

---

11   See: Jules Antoine Lissajous, "Mémoire sur l'étude optique du mouvement vibratoire", *Annales de Chimie et Physique* 2 (1857), pp. 147–230.

takes profit of a "budding guide" programme set up by the city itself. The visiting pupils have then access to an experimental laboratory to make simple experiments connected with the exhibition. We may note the spontaneous reaction of a pupil whom the teacher met at the market in Versailles: "Madam, thank you very much, I liked it very much and I repeated the experiment at home".

– High School students export their enthusiasm for the heritage of the high school museum. They welcome teachers and high school students from other schools in Versailles on visits to the museum.

– Each year, they also present the museum's objects to a group of middle school students in less favourable socio-economic sectors of the Mantes-la-Jolie city (in the so-called "Zone d'Education Prioritaire" (ZEP, Priority Education Sector)) in collaboration with the association *ConScience*, led by high school students from the Lycée Hoche and with the support of a science teacher.

– Finally, language and literature teachers are also involved in the project, since the museum also welcomes foreign exchange students of middle and high school levels. The preparation of French translations of historical device manuals, originally published in Italian, German or English, is part of a foreign language pedagogical activity that supports the integration of the foreign exchange students with the historical instruments.

In addition to these projects, which are regularly undertaken each school year, other, more specific, activities may emerge in a particular year. For instance, in 2014, the students represented France at the "Stockholm Junior Water Prize". In this international competition, 15-year-old students are tasked with developing school projects that address major water challenges. Supervised by teachers of Engineering Sciences and Earth Sciences and Life, secondary students made an Archimedean screw with the Lycée's 3D-printer, and used it in a model of a wastewater treatment plant.

The idea of the project emerged thanks to the Archimedes's screw made according to Nollet's instructions, in the possession of the Lycée museum (Fig. 7.5). Originally, this screw model, made of varnished and decorated wood, was designed to raise a ball from one height to another. The students learned that a similar device was already used by the Egyptians to raise water for agricultural purposes. The students in the second year of secondary school found how to combine this old tradition with the most recent techniques, such as 3D-printing. More recently, the whole school has been working to prepare a large exhibition entitled *Histoires de l'eau* (Fig. 7.6).[12] The students studied

---

12   *Histoires de l'eau: L'eau et Versailles du XVIIe au XXIe siècle* (Stories of Water: Water and Versailles from the 17th to the 21st Century), Musée du Lycée Hoche, Versailles, September 23, 2019–December 18, 2020.

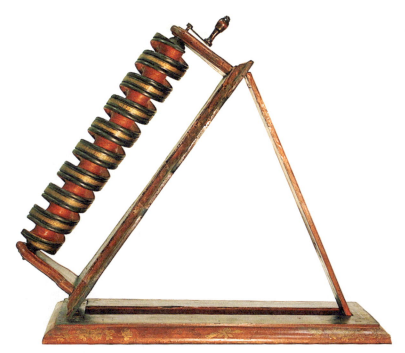

FIGURE 7.5 The Archimedes screw from the Lycée Hoche of Versailles
PHOTO JEAN MILLET

water from different points of view: chemistry, hydraulic engineering, arts, and so on. The exhibit, organised by the professors with the students, used all the historical devices devoted to water studies or hydraulics found in the Lycée's collection.

4 Project with University Level Students

As in secondary schools, some French universities have preserved their heritage of physical instruments, but contrary to the secondary school curriculum, the use of this heritage is absent from university curricula. In fact, in France, the secondary school curriculum is fixed at a national level by the Ministry of Education. On the other hand, universities have total independence to decide on their detailed curricula; nevertheless, a validation of each curriculum from the Ministry of Education is required. According to our sources, there is no inclusion of the practical use of historical instruments in the French university curricula, either in Physics or in History of Sciences. As far as we know, the project we present here pioneers the practical use of historical instruments in

FIGURE 7.6 Poster announcing the exhibition *Histoires de l'eau*, Musée du Lycée Hoche, Versailles, September 23, 2019–December 18, 2020

French university curricula, that is as part of a lecture unit, beyond internship work. This project concerns first year students of the Master's degree of Physics at the University Claude Bernard Lyon 1 (UCBL) and involves the restoration and augmentation of historical physics instruments from the heritage collection of that same university. The UCBL owns, in fact, a physics instruments heritage of about 200 objects from which 59 are already included in the ASEISTE directory. The project was first initiated in 2018 by two of the authors of this chapter, A. San Miguel and F. Khantine-Langlois, together with the student's club *Association des Physiciens de Lyon* (APL), and with the support of the *Société Française de Physique*. The project was proposed by us to a group of ten students, from the students' club, who then approached their professor, asking for the possibility to make a project in connection with physics and science outreach. In 2018, the project emerged as an out-of-curriculum activity of the students' club. In 2019, considering the success of the project, the enthusiasms of students and its other benefits, the project was proposed and accepted as optional laboratory practical work integrating the Master's degree of Physics curriculum in an "Experimental Tools" learning unit.

The objectives of the project are then numerous:
– to participate in the restoration (Fig. 7.7) and augmentation of historical physics instruments from the heritage collection of the UCBL;
– to promote the interest of students for history of science and instrumentation;
– to disseminate the scientific cultural heritage for a large audience.

Up to now, seven instruments of the UCBL collection have been involved in this project:
1. a Zénobe Gramme machine (Fig. 7.7 and Fig. 7.8, top left);[13]
2. a Morin free fall machine (Fig. 7.8, top left and right);[14]
3. an oscillatory fall gravimeter;[15]
4. a Deprez-D'Arsonval galvanometer acquired in 1943;[16]
5. a Bourbouze galvanometer (Fig. 7.8 bottom);[17]

---

13   On this machine see: Antoine Breguet, "La machine de Gramme", Gauthier-Villars, Paris, 1880.
14   See: Jules Salleron, *Notice sur les instruments de précision*, Salleron, Paris, 1864, pp. 21–22.
15   The gravimeter is similar to the one described in: Anonymous, "Etude de la chute libre. Exercices pratiques de physique, classes de mathématiques et de philosophie", *Bulletin de l'Union de Physiciens* 265 (1932), pp. 450–458.
16   See: Eric Gerard, *Mesures électriques*, Gauthier-Villars, Paris, 1912, pp. 155–191.
17   See: Jean-Gustave Bourbouze, "Galvanomètre vertical à fléau", *Journal de Physique Théorique et Appliquée 1*, 1 (1872), pp. 189–190.

FIGURE 7.7
Master students from University Claude Bernard Lyon 1 (UCBL) working on the restoration of a Zénobe Gramme machine

6. a Cotton balance;[18]
7. a Van de Graaff generator for teaching purposes (Fig. 7.8 top left).[19]

The last two instruments were both probably fabricated in the 1960s.

The students worked in groups of two or three on each instrument. In some cases, they needed to totally disassemble the instruments and carefully clean them, or treat the surfaces (see Fig. 7.7). When needed, lost or broken pieces were either acquired for replacement or produced at the university workshops, always taking care in preserving the general aspect of the instrument. We asked the students to produce a file describing the instrument (materials,

---

18 See: A. Cotton and G. Dupouy, *Comptes Rendus du Congrès International d'Electricité*, Gauthier-Villars, Paris, 1932: v. 3, p. 207.
19 See: R.J. Van de Graaff, K.T. Compton and L.C. Van Atta, "The Electrostatic Production of High Voltage for Nuclear Investigations", *Physical Review* 43 (1933), pp. 149–157.

USE IN EDUCATION OF HISTORICAL PHYSICS INSTRUMENTS   135

FIGURE 7.8 *From top left to bottom*: Master students from University Claude Bernard Lyon 1 (UCBL) presenting restored physics instruments at an exhibition, characterising the Morin machine, and characterising the Bourbouze galvanometer

dimensions) and information on its history and operations. For this, they studied the instrument itself and consulted both the web and library resources. An important part of the project included making the instrument work and characterising, when possible, its accuracy or dynamical range. We then provided the students with a simple portable recording studio including lighting spots, uniform backgrounds, stabiliser for recording with mobile phone and microphones. They were requested to produce at least two videos to be published on the University YouTube channel: a first one in which the instrument was presented and shown at work, and a second one in which the physical principles of the apparatus were described.[20] The students' work was evaluated considering both their investment and the quality of their productions. This project required a lot of dynamism both for the students and the professors. In fact, it was quite common to make a couple of visits to electronics or mechanics workshops during class time in order to try to solve some problems with the material with the help of technicians, or to borrow some instruments for measurements from other faculty members. This experience was much appreciated by the students who had really the feeling of being in a research situation rather than a typical laboratory practice, in which a preset protocol had to be reproduced. For instance, in the original Cotton balance project, mercury was used on one moving electrical connection. The use of mercury is currently forbidden in France academic teaching laboratories. The students had the idea of replacing it by the thinnest possible copper wire available in the electronics laboratory. They successfully applied their idea allowing the Cotton balance to work again while respecting today's safety rules. The mechanical workshop was also consulted for the adaptation of a pencil support to trace the parabolic fall in the Morin free fall machine. For this, the students themselves designed and produced the drawings.

The students were really involved in all the steps of the project, taking different initiatives, such as contacting a retired instrument maker, involving other students with competencies in video editing, or even applying for resources from museums, including the use of copyrighted images.

After validation by the professors, the students' productions were integrated in the historical physics instruments web page of the UCBL,[21] including links to the associated video playlist on YouTube. Some of the instruments are exhibited in the hall of the main physics building of the UCBL, with QR codes to enable viewing of the students' videos. In some cases, the students also prepared a quiz, either as a questionnaire or in video format, which is also made available to the visitors. A link to the UCBL YouTube channel referred above can be found in the ASEISTE web page.

We had very good feedback from students, with some of them spontaneously expressing that they really liked the experience a lot. The interest of the students in this pedagogical approach appears clearly when considering the number of answers to a mail survey which we proposed during the 2020 summer holiday period. The survey included both Likert scale items and open

---

20 The videos are available at: www.youtube.com/playlist?app=desktop&list=PL_ZZHW rfptdSQc3wsLvE1nGeKQH8c93Lg (accessed 24 Oct. 2020).

21 https://collectionsphysique.univ-lyon1.fr/ (accessed 24 Oct. 2020).

questions. A total of eighteen students participated in the project, including ten during the 2018–2019 academic year and eight in 2019–2020. A total of ten respondents to the survey were received. The survey objective was to better characterise the impact of the project on the student's interest in history of science or instrumentation. First we wanted to know the students' motivations for having chosen to participate in the project on historical physics instruments. All students were interested either in history of sciences or in science outreach, or both. 80% of the respondents also declared that, after having participated in the project, they experienced personal increase in their interest in history of sciences or in scientific instrumentation. We can then conclude that the approach of the pedagogical project here presented, contributes significantly to improve the interest of students for the history of science and scientific instrumentation, but that such an impact applies to a student population already showing an inclination for these subjects.

It is still too soon to evaluate the impact of the project for non-participating students or visitors. In fact, the students' productions are accessible through QR-codes in the exhibited instruments at the hall of the physics department building at the UCBL for only a few months. Every forthcoming academic year, we plan to involve students in conducting similar developments for four additional instruments, or for an experimental set-up based on an instrument not previously represented in the programme.

5 Conclusions and Prospects

In France, a particular network including secondary schools and universities has emerged through the ASEISTE, which contributes to the preservation and knowledge of an important heritage of historical physics instruments, most of them originally conceived for teaching purposes. We presented a number of initiatives at secondary school and university level, which have fostered fruitful projects in which historical physics instruments participate in pedagogical undertakings. In our examples a historical, scientific and technical heritage combines productively with modern creative tools (3D-printing) or communication approaches.

To further develop this type of initiative, we are considering different paths for the future:

– Development through emulation. We plan to present the projects described here in national and international congresses on innovative pedagogy. Our presentations may stimulate analogous projects, or further the evolution of the ones presented here, creating synergic effects in association with the ASEISTE;

- Development through collaboration. ASEISTE can contribute to articulate initiatives that bring projects involving historical instruments to other secondary schools or universities which do not possess such heritage. The possibility of developing 3D-scans of original instruments, or of applying 3D-printing as tool to make parts, may be a basis for collaborative initiatives between secondary schools. Putting students of two different schools or universities in such types of collaboration, including between foreign countries, offers enormous potential.
- In the university projects, we plan to produce new videos showing the restoration work done by the students themselves.

All these projects participate in many aspects of the preservation and use of scientific instruments for education. We can first consider the direct outcomes of the projects as developing resources (videos, files, exhibitions) which contribute to better document scientific objects or to spread their knowledge to a general public. Once our project was initiated, we were then in a position to propose collaborations with institutions such as universities, research bodies or the Education Ministry, and learned societies such as the *Société Française de Physique*. By partnering with these groups, we can now participate in subsequent developments having different methods of support, including funding.

Finally, our present experience with the use of the heritage of scientific instruments as part of pedagogical projects, is very stimulating both for students at all ages – from primary school to university levels – and for the professors. In particular, we have found that historical scientific instruments are an excellent means of developing networked projects between different disciplines.

CHAPTER 8

The Use of the Museum Collection for Educational Purposes

Roland Carchon and Danny Segers | ORCID: 0000-0002-9019-2274; 0000-0002-5782-7477

1   Introduction

Museum collections at a university-related museum are mainly build up with materials that have a link with the educational aspect and research of the past. This is particularly true for scientific instruments that relate to diverse fields of sciences. These collections have, in general, a high degree of specialisation, that often limits the accessibility and interest of the general public. Nevertheless, these historical objects were fundamental to obtaining the actual knowledge of the various fields and often integrate considerable intellect combined with instrument makers' skills.

This paper will give an overview of the educational needs and museum possibilities in *Section 2*. The use of part of the collection of scientific instruments for educational purposes, being bachelor (undergraduate) and master projects, PhD work, and lectures for high school, is given in some detail in *Section 3*. Some of the work mentioned initiated proper intern-museum research on instruments, as given in *Section 4*. A conclusion is given in *Section 5*.

The activities described in this paper relate to the former Museum for the History of Sciences of the University of Ghent (UGhent). From 2018, it was decided to integrate this into the new Ghent University Museum (GUM) that consists of six former partial collections, comprising ethnography, archaeology, zoology, morphology, history of sciences and history of medicine. The text in this chapter only discusses the Collection for the History of Sciences and no extrapolations from it should be made about the other collections.

The former Museum for the History of Sciences had a very elaborate collection of physics instruments. Besides the usual items (such as electrostatic generators, Leyden jars, galvanometers, microscopes, and so on) the museum also had some unique sub-collections. These are centred around two important scientists who worked at the university: Joseph Plateau (1801–1883) and Leo Baekeland (1863–1944).

© ROLAND CARCHON AND DANNY SEGERS, 2022 | DOI:10.1163/9789004499676_010

What follows refers to the field of physics, which is the authors' own discipline.[1]

Relating to academic heritage, two attitudes can exist: one that prohibits any interference with an object for reasons of "authenticity", and one that would allow restoration. That objects can be restored without any risk and without losing historical value has been demonstrated with the Tartu (Estonia) telescope.[2] Only those topics were considered for bachelor projects where the use of such objects was allowed.

2   Educational Needs and Museum Possibilities

The educational use of collection items responded to needs from the academic world, non-university higher education and high schools. At UGhent, the bachelor in physics education is a three-year programme. Each year has a quotation of 60 European Credit Transfer System (ECTS) credits. Most of the topics are theoretical courses in the field of general physics, theoretical physics, mathematics, astronomy, computer science and materials science. Each year there is also one course on "Experimental Physics and Astronomy; Data Reduction" where practical exercises are carried out. These are classroom experiments to be reproduced by the students.

The third year bachelor course in the physics and astronomy curriculum contains a topic called the "Bachelor Project", that has a quotation of six ECTS-credits, the same as the other topics. This bachelor project has the following requirements: 1. it should be "independent" experimental work in the physics and astronomy domain; 2. it takes place during the third year, second semester and has a minimum of twelve half days as contact hours; 3. it should be documented in a thesis of some fifteen pages (according to the instructions given in the course on experimental physics), and be presented to a public audience at the end of the semester during a twenty-minute session, questions included. Due to the time limitation for the overall bachelor project, some support from museum staff members was needed.

The museum contributed by accepting two students per year and guiding them to perform the experiments, to write the report and to give the

---

1   This paper gives an update and extension of the presentation: R. Carchon, D. Segers, K. Wautier and A. Jonckheere, "The Use of the Museum Collection for Educational Purposes", *XXXIII Symposium of the Scientific Instrument Commission*, Tartu, Estonia, 25–29 August 2014.

2   Paolo Brenni and Ileana Chinnici, "From Palermo to Tartu: restoring, preserving, displaying Merz telescopes", *XXXIII Symposium of the Scientific Instrument Commission*, Tartu, Estonia, 25–29 August 2014.

THE USE OF THE MUSEUM COLLECTION FOR EDUCATIONAL PURPOSES   141

presentation. The topics are defined by the museum according to whatever is possible and introduced to the project committee for further allocation to the students to make their choice. If possible, experimental results obtained by modern means are compared with the historic (authentic) measurements.

A similar requirement exists for the master's thesis, which is more voluminous (50 to 100 pages) and involves more workload (30 ECTS-credits). We do not recommend undertaking a master's thesis in the museum, unless the student is looking for a career as a high school teacher.

PhD work is only possible in the field of history, where the museum only acts as sub-contractor and technical support.

We support teaching and demonstrations for groups of students who are in their final year of high school. The workshops are in general initiated during the temporary exhibitions that change with the start of an academic year. In this way every year a new package is offered, which normally continues to remain available in the museum's offerings to schools.

3 The Different Projects for University and High School Students

In this section, we give an overview of the different education activities related to the museum: bachelor and master projects at the UGhent, Physics & Astronomy department; PhD work at the UGhent, Philosophy department; bachelor projects at non-university higher education institutes; workshops for high-schools and the course UGhent, Free University of Brussels (VUB) on the history of sciences.

3.1 Bachelor Projects (University Level)

In *Table 8.1*, a summary is given for the different bachelor projects realised in the museum. The required skills for the students and specific benefits for the museum are indicated. The first column code relates each thesis subject to the corresponding course unit of the discipline for Bachelor in Physics and Astronomy.

In general, the bachelor projects are complementary to the theoretical courses and the students have to learn to solve problems in an autonomous way. Specifically, for bachelor projects with museum objects, the historical background and context of the instruments invites the student to consider the human beings that pursue the work of science, to learn their stories and begin to identify with everything that those human beings faced, such as physical, technical, and community issues. Moreover, there is a link to a course in basic physics.

The museum benefits from the bachelor student projects by testing the working state of the instruments and their performance. As a result of the student projects, the museum collection is more thoroughly examined and characterised.

The two projects with the Plateau collection are described in more detail below.

TABLE 8.1 Bachelor projects realised in the University of Ghent (UGhent) Museum. In the first column the different course disciplines are indicated by the numbers within round brackets: (1) General Physics: Electricity and Magnetism, (2) General Physics: Waves and Optics, (3) General Physics: Quantum Mechanics, (4) Materials Science: Atom and Molecular Physics, (5) Materials Science: Nuclear Physics, (6) Materials Science: Particle Physics, (7) Experimental physics and data reduction

Course discipline	Student skills	Benefits for the museum / collection
Two projects with the collection of Joseph Plateau		
a)	Construction of a modern version of the anorthoscope	
b)	A replica of the bioscope of Duboscq	
(2), (7)	– construction of an instrument from scratch – Instrument control	– Demonstration setup for the visitor
Six Projects on electricity		
a)	electrostatics	
b)	electric discharges in gases	
c)	galvanometers	
d)	capacity of Leyden jars	
e)	characteristics of alternating currents	
f)	microwaves	
(1), (7)	– practice with modern instruments – principle of galvanometer – principle of capacitor – frequency changes in alternating currents – microwave generation	– working status of equipment – application in workshop and demonstrations
Two projects on acoustics		
a)	vibrating strings and their use in music instruments	
b)	acoustic resonators	

THE USE OF THE MUSEUM COLLECTION FOR EDUCATIONAL PURPOSES   143

TABLE 8.1   Bachelor projects realised in the University of Ghent (UGhent) Museum (*cont.*)

Course discipline	Student skills	Benefits for the museum / collection
(2), (7)	– understanding waves – application of waves to music instruments – use of resonators	– information for workshop – frequencies of resonators known

Three projects on optics
    a)   diffraction
    b)   microscopy
    c) small lenses and their magnification		

(2), (7)	– Diffraction patterns explored – lens manufacture by grinding and blowing	– van Leeuwenhoek originality confirmed

Two projects on nuclear physics
    a)   cloud chamber
    b) e/m of the electron and Thomson's experiment		

(5), (6), (7)	– complexity of cloud chamber – precision of Thomson results	– cloud chamber and Thomson's experiment can be demonstrated

3.1.1   Construction of a Modern Version of the Anorthoscope of
            Joseph Plateau

Joseph Plateau organised his teaching based on experimental demonstrations.[3] Therefore he set up a collection of physics demonstration instruments which are still an important collection in our museum, now known as "The cabinet of physics". This approach was very revolutionary at this time. Plateau researched moving images and the persistence of vision. He is still recognised as father of film. Around 1830 he developed an 'anorthoscope': a device that uses rotation to deconvolute (or unroll) an anamorphised object. In general, an anamorphised object is a distorted image which appears normal only when observed either from a particular point of view (perspective and plane anamorphs) or with some optical device (curvature mirror, anorthoscope with rotational motion, and so on).

---

3   Kristel Wautier, Alexander Jonckheere and Danny Segers, "The Life and Work of Joseph Plateau: Father of Film and Discoverer of Surface Tension", *Physics in Perspective* 14 (2012), pp. 258–278.

FIGURE 8.1 Anorthoscope disk of Plateau
COURTESY GHENT UNIVERSITY MUSEUM

We have a number of original anorthoscope devices and discs of Joseph Plateau in our collection (Fig. 8.1). On the device, two discs are mounted in front of each other: in the back the disc with the deformed image and in the front a disc with a number of slits (for example, four). The two discs are mounted so as to revolve in opposite directions with a constant ratio in angular speed (that is, $\omega_{\text{deformed image}} = 4\,\omega_{\text{slits}}$). In Plateau's design this is achieved with a mechanical gear transmission. This is not very versatile because the angular speed ratio cannot be changed. While both discs are counter-revolving, the viewer looks through the disc with the slits, and glimpses the deformed image on the disc behind it. When the ratio of the angular speeds is correct and constant, the deformed image resolves for the viewer into a "normal" or identifiable form.

THE USE OF THE MUSEUM COLLECTION FOR EDUCATIONAL PURPOSES 145

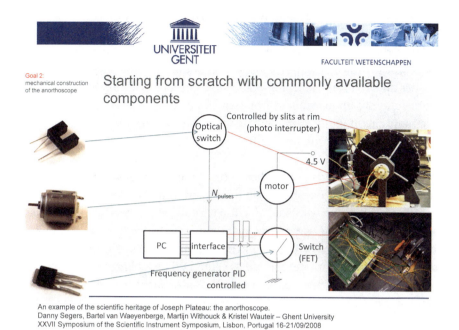

FIGURE 8.2 Slide explaining the functioning of the anorthoscope of Joseph Plateau
COURTESY GHENT UNIVERSITY MUSEUM

A student initiated a bachelor project based on this historical instrument by constructing a modern version where the angular speed of the discs is controlled with electronic circuits. This offered the possibility for the viewer to change the angular speed ratio (Fig. 8.2).[4]

The electro motors were steered by a circuit with a frequency generator, proportional, integral and differential (PID) controller and field effect transistor (FET) switch. The angular speed of the discs was measured with a photo interrupter. The electronic circuits were computer-controlled by a programme written in LabView. The student project produced a working modern instrument where copies of the original historical discs could be used to view and discern the undistorted images. The student tested the instrument by producing a disc

---

4 Danny Segers, Bartel Van Waeyenberge and Martijn Withouck, "An Example of the Scientific Heritage of Joseph Plateau: The Anorthoscope", *XXVII Symposium of Scientific Instrument Commission*, Lisbon, Portugal, 16–21 September 2008.

with a distorted image of a perfect circle. The mathematics as described by Hunt, Nickel and Gigault was used.[5]

This project resulted in a number of skills that the student developed: *i.* study the historical research of Joseph Plateau; *ii.* study the physics and mathematics underlying anamorphoses and anorthoscope discs; *iii.* the mechanical construction of an instrument from scratch; *iv.* the construction of a simple electronic circuit for measurement and control; *v.* learn to work with LabView to control interface cards, PID controllers, and so on.

3.1.2 Construction of a Replica of the Bioscope of Jules Duboscq

Another device that Joseph Plateau developed in his research on the persistence of vision in 1832 consisted of a disc with sixteen sectors and sixteen slits at the rim.[6] In each sector, a drawing of a moving object is placed in one of sixteen phases that make up the total movement. The disc has to be mounted towards a mirror and rotates around its centre. When looking through the slits to the mirror images, the observer sees the subsequent flashes of the different phases of the moving object. If the rotation speed is high enough, the eye will not see the individual images anymore: due to the persistence of vision, the illusion of a continuous motion is created. The first cartoon film was produced with this type of device and that is why Plateau is known as the father of film. This disc and the corresponding apparatus became later known as the phenakistiscope.[7]

In France, around 1851, the instrument maker Jules Duboscq (1817–1886)[8] built a new instrument called the bioscope, that combines the principle of the stereoscope[9] with the generation of moving images of the phenakistiscope of Plateau. He used stereoscopic photographs and created the first 3D film. The bioscope was commercialised around 1856. To our knowledge no remaining

---

5 The mathematics of anamorphic images was recently described in J.L. Hunt, B.G. Nickel and Christian Gigault, "Anamorphic Images", *American Journal of Physics* 68 (2000), pp. 232–237. The functioning of the anorthoscope was also recently re-investigated in J.L. Hunt, "The Roget Illusion: The Anorthoscope and the Persistence of Vision", *American Journal of Physics* 71 (2003), pp. 774–777.

6 Joseph Plateau, "Sur un nouveau genre d'illusion d'optique", *Correspondance Mathématique et Physique* 7 (1832), p. 365.

7 J. Plateau, "Des illusions optiques sur lesquelles se fonde le petit appareil appelé récemment Phénakisticope", *Annales de Chimie et de Physique de Paris* 48 (1833), p. 28.

8 P. Brenni, "19th Century French Scientific Instrument Makers – XIII: Soleil, Duboscq, and their Successors", *Bulletin of the Scientific Instrument Society* 51 (1996), pp. 7–16.

9 Charles Wheatstone (1802–1875) studied in the same period binocular vision. In the period 1830–1832 he developed the first stereoscope. He published the results in: C. Wheatstone, "Contributions to the Physiology of Vision – Part the First: On Some Remarkable, and Hitherto Unobserved, Phenomena of Binocular Vision", *Philosophical Transactions* 128 (1838), pp. 371–394.

FIGURE 8.3 One-page leaflet on the bioscope, made by Duboscq for commercial reasons, printed in Paris in 1853
COURTESY GHENT UNIVERSITY MUSEUM

bioscope still exists today. From the French patent of Duboscq of 1852 we have an idea of how the apparatus looked (Fig. 8.3).[10]

10   Jules Duboscq, *Brevet d'Invention S.G.D.G., No. 13 069, troisième certificat d'addition*, Paris, November 12, 1852.

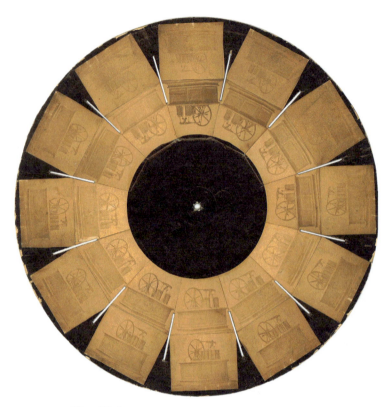

FIGURE 8.4 Disc of Duboscq
COURTESY GHENT UNIVERSITY MUSEUM

The GUM possesses a unique bioscope disc of Duboscq with stereoscopic photographs of a steam engine. To our knowledge, it is the only existing Duboscq disc (Fig. 8.4).[11] It consists of twelve sectors with a radial slit near the rim of each sector. In each sector, two stereoscopic photographs are placed, one above the other. The photograph for the right eye is placed at the rim of the disc and the photograph for the left eye is placed more to the centre of the disc. In each of the twelve sectors are placed successive photographs through the motion cycle. The disc is mounted on the apparatus and can be rotated. Two mirrors are facing the images. The observer looks from behind the disc through two adjacent slits at the reflected images of the mirrors. The image for the left eye is projected on the leftmost slit, while the image for the right eye is projected on the rightmost slit. When the disc rotates, the observer sees a stereoscopic moving film.

11   Maurice F. Dorikens, *Joseph Plateau 1801–1883. Leven tussen kunst en wetenschap; Living between Art and Science*, Province of East-Flanders, Ghent, 2001, p. 49.

THE USE OF THE MUSEUM COLLECTION FOR EDUCATIONAL PURPOSES 149

FIGURE 8.5
Replica of bioscope of
Duboscq
COURTESY GHENT
UNIVERSITY MUSEUM

The aim of this bachelor project consisted in making a replica of the bioscope of Duboscq. Due to the fragility of the original Duboscq disc it could not of course be used in this study. So the student had to study the historical context and prepare a new Duboscq disc with homemade stereoscopic photographs of a rotating model of a flower. The student was also actively involved in the construction of the replica (Fig. 8.5).[12] It was a very interesting project where

---

12   Katrien Meert, *De Bioscoopschijf van Duboscq*, Bachelor thesis, University of Ghent, academic year 2008–2009.

the student had to make inferences and decisions without a direct model to work from.

3.2 *Master Dissertations*

3.2.1 Study of the "Cabinet of Physics" of Joseph Plateau

Joseph Plateau built up an important collection of demonstration instruments, now held at GUM and entitled "the cabinet of physics of Plateau".

For a master's thesis, one student examined and characterised Plateau's cabinet of physics.[13] The framework of this thesis examines the instruments for their operational status and vulnerability toward potential damage and the possibility of minor repair. This framework was used to select some instruments including: *i.* instruments for the study of fluids and Pascal's law, *ii.* Heron's fountain, and *iii.* Archimedes' principle and the hydrostatic balance.

In the thesis project, the selected instruments were made operational for use in the same demonstration experiments as in Plateau's teaching. In the central library of the university, lecture notes made by three students during Plateau's time teaching were available. These copies revealed that the historical students had a very precise idea of the demonstrations given by Plateau. These historical lecture notes were invaluable to the master's student in undertaking the use of the very same instruments to perform the demonstrations and to film these demonstrations for sharing with the public. Our master's student filmed a version of these demonstration experiments and mounted the film on an iPad so that visitors to the museum could see real demonstrations with some objects of Plateau's cabinet. This was appreciated by the visitors. The instruments used in these demonstrations were the original Plateau apparatus, as repaired and made operational by the master's student.

In the thesis, the physics principles behind those experiments were described in a historical context and at the end of each chapter, a student project for high school physics teaching was added.

3.2.2 Reproduction of Melloni's Historical Experiments on the Nature of Infrared Radiation

The Melloni bench belonged to the Plateau cabinet (Fig. 8.6). A recapitulation of Melloni's experiments on infrared light was chosen as the subject of a master's thesis, extended to modern physics theories.[14]

---

13   Martijn Withouck, *Bestudering 'Cabinet de Physique' van Joseph Plateau*, Master thesis, University of Ghent, academic year 2009–2010.

14   Sien Cromphout, *Mijlpalen in het onderzoek naar Warmtestraling – de bank van Melloni en infrarood metingen van zwarte stralerspectra*, Master thesis, University of Ghent, academic year 2010–2011.

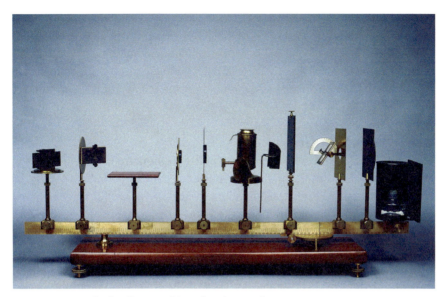

FIGURE 8.6 Bench of Melloni, used for infrared research
COURTESY GHENT UNIVERSITY MUSEUM

The radiation sources that Macedonio Melloni (1798–1854) used were based on: *i.* heating with a small alcohol flame (a hot metal plate, a glowing platinum wire, and a lamp of Locatelli) or *ii.* the radiation produced by a small copper cube (6 cm size) filled with boiling water (known as the Leslie cube). The Leslie cube was reconstructed by the master's student, and the platinum wire was changed after each experimental trial that the student performed. The measurements consisted in detecting a small signal in a high environmental background, which is, obviously, always difficult for any experiment. Signal detection was through a modern thermocouple and digital voltmeter.

The opportunity was taken to have a full characterisation of the variety of original absorbers, by using IR-UV spectrophotometry, Fourier transform infrared spectroscopy (FTIR), X-ray diffraction and secondary electron microscopy (SEM). Before this master's thesis, the museum had no precise characterisation of the absorbers that were available. So this characterisation effort of the student project also helped in completing the museum inventory.

The study of Melloni's historical experiments on infrared radiation was continued by the student, focusing more on the importance of Melloni in the history of physics. The master's student investigated in detail how Melloni came to his conclusion that heat radiation is of the same nature as light. By contrast, his fellow scientists William Herschel (1738–1822) and Henry Draper (1837–1882) did not regard heat as a form of light radiation.

A literature study by the student established that Melloni did experiments and published on the polarisation of heat radiation.[15] This seems a rather decisive argument: indeed polarisation works for light, but not for acoustic waves, a topic well known at the time. Moreover, the fellow scientists who did not accept Melloni's perspective on heat also did not perform polarisation experiments.[16]

3.3 *PhD Dissertations*

3.3.1 PhD Dissertation on Leo Baekeland

Leo Baekeland was born in Ghent. He studied at Ghent university (1880–1884) and obtained the degree of doctor in science. In the period 1885–1889 he became assistant of Professor Théodore Swarts (1839–1911) at the chemistry laboratory. In 1889 he emigrated to the US where he invented a new photographic paper, Velox, and in 1907 the first plastic, Bakelite.

The GUM holds an extensive collection relating to Leo Baekeland. At Ghent University he performed research on photo-chemistry. A number of crystals prepared by him for his research are part of the collection. From the period when he stayed in the United States, the museum has a huge number of correspondence letters and photographs. Soon after his invention of Bakelite, Leo Baekeland prepared and sent three objects to Professor Frédéric Swarts (1866–1940), the successor of Théodore Swarts (Fig. 8.7). These objects are still present in the collection of the GUM.

Frédéric Swarts performed research on fluorine-related compounds. These compounds are very chemically active and aggressive. Bakelite is chemically inert and is ideal for working with all kind of aggressive chemicals, such as those that Frédéric Swarts researched.

In 2007 Ghent University celebrated the centenary of the invention of Bakelite, by organising a 'Baekeland year'. On that occasion an exhibition on Baekeland and Bakelite was set up by the Museum for the History of Science of Ghent University.

Afterwards the Museum for the History of Science and the department of History collaborated in developing a common research programme, resulting in a doctoral thesis on Leo Baekeland.[17] During this study, the historical

---

15   Macedonio Melloni, "Mémoire sur la polarisation de la chaleur", *Annales de Chimie et de Physique* 61 (1836), pp. 375–410.

16   R. Carchon and D. Segers, "Melloni's experiments and the search for the nature of infrared radiation", *European Journal of Physics Education* 6, 4 (2016), pp. 12–26.

17   Joris Mercelis, Leo H. Baekeland (1863–1944) as Scientific Entrepreneur: A Transatlantic Perspective on the Science-Industry Nexus, PhD thesis, UGhent, academic year 2012–2013; subsequently published as J. Mercelis, *Beyond Bakelite: Leo Baekeland and the Business of Science and Invention*, MIT Press, Cambridge (MA), 2020.

FIGURE 8.7 One of the three objects prepared by Leo Baekeland for Professor Frédéric Swarts
COURTESY GHENT UNIVERSITY MUSEUM

documents of Baekeland, present in the museum, were extensively used. This historical study investigated the relationship between (natural) science and industry and their mutual dependence. In the framework of this research, the intellectual ownership (patents, trademarks, copyright) and the scientific entrepreneurship were studied in a transnational context.

3.4 *Bachelor Projects (Non-University Level)*
The Museum for the History of Science has facilitated a number of bachelor projects for bachelor students of a teacher training college. The subjects were originally proposed by the teacher training college itself and the museum collection was put at the disposal of the students to work out their projects.

3.5 *Workshops for High-School (16–18 Years Old)*
The Museum for the History of Science has organised specific workshops for high-school pupils. The following subjects were treated: science in the kitchen, a short history of electricity with interactive demonstrations, vacuum and space, music and resonance, and cartography and geomatics.

The workshops were organised as lectures mainly consisting of experiments, with pupil participation. The teacher could interrupt and make connections with classroom teaching, substantially increasing the effectiveness of the workshop.

3.6 Course "Elements of the History of Science" by UGhent and VUB

In the period from 2006 till 2016, a common course on "Elements of the History of Science" was organised by the former Museum for the History of Sciences – UGhent and the Free University of Brussels (VUB).

The course was an optional choice for the second year master's students in the faculty of science of both universities. It had a quotation of 6 ECTS-credits. The course was divided into two parts: formal lectures and practical exercises.

The formal lectures on the history of physics and astronomy were organised by the VUB. Sometimes this was complemented with guest lectures on contemporary topics.

The part on practical exercises was organised by the Museum for the History of Sciences. It consisted of two parts:

i. The students had to read an historical book and write a summary according to "Instructions for Authors" of a certain scientific A1 journal. They also had to prepare an oral presentation for their fellow students. To facilitate this part of the course a few lessons were given on "scientific communication".

ii. The students had to give a guided tour on a well-chosen part of the collection. Therefore, they had to study the objects and the historical context of that part of the collection.

4 Scientific Research of Museum Objects

Some of the educative projects gave rise to museum-initiated research of collection objects. The study concerning Melloni's bench has been already mentioned, and others included the construction of a replica of a Huygens vacuum pump, an armillary sphere and an investigation of the origin of the van Leeuwenhoek microscope. The latter has a link with bachelor projects.

4.1 The Origin of the van Leeuwenhoek Microscope

The bachelor thesis on microscopy initiated the question of whether the brass microscope that we have in the collection is original or a copy from a later time period (Fig. 8.8).[18]

---

18   In fact, the microscope belongs to the van Heurck collection of the city of Antwerp. It is on permanent loan to the University of Ghent, where it is under the responsibility of the GUM, collection history of sciences.

THE USE OF THE MUSEUM COLLECTION FOR EDUCATIONAL PURPOSES   155

FIGURE 8.8
The original van Leeuwenhoek microscope from the collection "Stad Antwerpen"
COURTESY GHENT UNIVERSITY MUSEUM

Antoni van Leeuwenhoek (1635–1723) did not give an account of his lens making capability, but claimed this was done by grinding and polishing, not by blowing.[19] We applied both these ways of lens manufacturing, thereby considerably increasing the student's understanding of microscope optics.

The determination of whether the microscope in the museum collection is an original was ascertained by non-destructive element analysis of the composition of its brass parts. Element analysers are now available in hand-held format; one such instrument was used to analyse and identify the fluorescence lines after the Leeuwenhoek microscope was irradiated with X-rays.

---

19   J. van Zuylen, "The Microscopes of Antoni van Leeuwenhoek", *Journal of Microscopy* 121, 3 (1981), pp. 309–328; Marian Fournier, "De doos van Pandora", *Gewina* 25 (2002), pp. 70–74.

We analysed the GUM microscope and a more recent replica. This analysis showed that the museum instrument exhibits element concentrations that are consistent with other brass instruments dating from van Leeuwenhoek's time. Therefore, our instrument qualifies as an original from that same era.

Apart from trace elements and impurities, we focused on copper and zinc concentrations, the main components that form together more than 99% of the elements present in the microscope. The concentrations are approximately 70% copper and 30% zinc.[20] It should be mentioned that in the time period after van Leeuwenhoek, the concentrations were varied slightly: 73% and 27% for copper and zinc respectively.[21]

An additional assurance of the GUM microscope being genuine resulted from the analysis of the thread of the screw, that the student performed. Standardisation of screw threading was introduced much later.

5   Conclusion

Recent times have shown an increasing interest in science objects that are no longer used and that are left behind in some storage area. Originally these instruments served as demonstration material for university courses or were used in research. Today these collections are regarded as "academic heritage".

The academic heritage of the University of Ghent is collected in its museum, the GUM. The staff members of the Collection for the History of Sciences, now part of the GUM, have taken the initiative to respond positively to a request for support in matters of the "Bachelor Project" in the curriculum of physics and astronomy, being experimental work. Where possible, collection items were used directly in these student projects. It is obvious that one should not touch rare and valuable original objects. The decisions regarding whether certain objects can be considered for use in an educational context and whether museum collections have to be treated in a dynamic or passive way, have to be made on a case-by-case basis.

In very general terms, the student learned how to handle a problem, how to solve it, and how to report on it. In parallel, the museum was helped with the characterisation of its collection, and with a detailed description of the functionality of the items.

---

20   Karl Hachenberg, "Brass in Central European Instrument-Making from the 16th through the 18th Centuries", *Historic Brass Society Journal* 4 (1992), pp. 229–252.

21   Tiemen Cocquyt, "De identificatie van een zilveren microscoopje van Antoni van Leeuwenhoek", *Studium* 4 (2015), pp. 198–211, and private communication.

A positive point in favour of the museum is the use of old instruments that display a direct link to the theoretical courses that students take as requirements. By contrast, the equipment used by all other physics laboratories of the university that contribute to bachelor and master projects, is exclusively modern and sophisticated. Very often, the student sees no direct link between this laboratory equipment and what is discussed in the theoretical courses. The modern equipment is mostly integrated and miniaturised and gives no immediate visible link to what is happening. The old instruments from the museum collection are "simpler" in comparison to modern equipment. The essence of an experiment is more transparent when the student can discern how the instrument works.

The historic context also invites the student to have a look at the "person behind the scientist", as was the motto of George Sarton of Harvard University.[22] The historical items help to understand the evolution of scientific ideas and developments up to today, as a continuous effort.

The thesis evaluation and presentation in front of the jury and fellow students has been positive over the past ten years. Students who wrote their thesis using museum resources have got good marks and achieved the top 10% of the bachelor cohort.

The main activities as described here concentrate on the physics collection, but similar activities exist for the zoology collection.

22   See: www.sartonchair.ugent.be/en/sarton/biography (accessed 9 Dec. 2020).

CHAPTER 9

Historical Scientific Instruments in Exploratory Teaching and Learning

Elizabeth Cavicchi | ORCID: 0000-0002-4265-1296

1 Introduction

Learners engage with historical scientific instruments in many of my presentations at Scientific Instrument Commission (SIC) Symposia.[1] The principal context for my educational studies is a university seminar that I have developed and taught for fifteen years at MIT's Edgerton Center, titled "Recreate Experiments from History: Inform the Future with the Past".[2] Sessions are evolving and open-ended, including observing outdoors and with materials; sharing observations; discussing questions and curiosities, responding to readings and historical resources, and collaborating with classmates. Field trips to museums, rare books libraries and other sites provide opportunities in which students examine historical materials directly, raising their own questions and insights. This seminar is offered for academic credit outside of any degree programme or requirement. There are no formal entry qualifications. Undergraduate and graduate students with any background may enroll, including cross-registering students from Harvard. The class size is intimate, typically

---

1 This chapter expands upon the following presentations: Elizabeth Cavicchi, "Conflict and Balance: Classroom Explorations with Historical Instruments, Science, Geometry, and More", *XXXIV Scientific Instrument Symposium*, Turin, Italy, 7–11 September 2015; E. Cavicchi, "Old Instruments Give Rise to New Explorations in Learning and Teaching", *XXXIII Scientific Instrument Symposium*, Tartu, Estonia, 25–29 August 2014; E. Cavicchi, "Stepping into the Past to Understand Time: Explorations with Astrolabes, Clocks, and Observation", *XXXI Scientific Instrument Symposium*, Rio de Janeiro, Brazil, 8–14 October 2012; E. Cavicchi, "Telescopes and Telescopic Acts Bring Galileo into my Classroom", *XXIX Scientific Instrument Symposium*, Florence, Italy, 4–9 October 2010; E. Cavicchi, "Reconstructing the Camera Obscura Effect: Becoming Optical Experimenters", *XXVII Scientific Instrument Symposium*, Lisbon, Portugal, 16–21 September 2008; E. Cavicchi, "Recreating the Bead Lens of a Seventeenth Century Simple Microscope", *XXVI Scientific Instrument Symposium*, Cambridge, MA, USA, 6–11 September 2007.
2 MIT OpenCourseware posts assignments, student work, and readings from the 2010 winter term of this seminar: https://ocw.mit.edu/courses/edgerton-center/ec-050-recreate -experiments-from-history-inform-the-future-from-the-past-galileo-january-iap-2010/ (accessed 13 Oct. 2020).

© ELIZABETH CAVICCHI, 2022 | DOI:10.1163/9789004499676_011

fewer than seven, or even two members. Between sessions, students observe; keep a journal; do readings; and complete the course by writing a reflective paper. Assignments suggest diverse options for observing, reading and reflection; students may select and follow personal and collective interests. Potential for academic stress is lowered through: building experiences together; my flexibility in responding to where students are; and the course being half of the credits of standard offerings.

Being together, the students and I form experiences: with the natural and social world, each other, historical figures, others of our time and in the future. Our lived experiences are openings to dialogue, experiment, question, and reflection. By taking action, such as observing the night sky or experimenting with lenses, students encounter something other, that may be intriguing and unknown for them, as it was for those in the past. Welcoming experience – however that evolves – is a mutual process of growing in respect, flexibility, spontaneity, and continuity. Understanding evolves; uncertainty abounds; creativity and new perspectives arise; evidence is produced and questioned; social concerns, history, reflection, and subjectivity are embraced. These qualities, inherent in our class experiences, are consonant with how many science educators today characterise the "nature of science".[3] Whereas those educators advocate for "explicit" instruction on each of these characterisations, my students directly access, discuss, and reflect upon the complex, contingent, and interactive relationships among each other, others, science, and the world.

As the teacher, I look to bring about experiences where students are explorers, and to sustain and extend their investigations.[4] In doing so, I play multiple roles – including researcher, participant, learner, advocate, and materials provider – throughout the experiences, practicing the pedagogy of "critical exploration in the classroom"[5] developed by Eleanor Duckworth from origins

---

3 National Science Teacher Association, Nature of Science, 2020, www.nsta.org/nstas-official-positions/nature-science (accessed 13 Oct. 2020).

4 See: E. Cavicchi, "Becoming curious science investigators through recreating with history and philosophy," in A.P.B. da Silva and B.A. Moura (eds.), *Objetivos Humanísticos, Conteúdos Científicos: Contribuções da História e da Filosofia da Sciência para o Ensino de Ciências*, Edupb, Campina Grande-PB, 2019, pp. 265–284, http://eduepb.uepb.edu.br/e-books/ (accessed 13 Oct. 2020); E. Cavicchi, "Learning Science as Explorers: Historical Resonances, Inventive Instruments, Evolving Community", *Interchange* 45, 3 (2014), pp. 185–204; E. Cavicchi, "Opening Possibilities in Experimental Science and its History: Critical Explorations with Pendulums and Singing Tubes", *Interchange* 39 (2008), pp. 415–442; E. Cavicchi, "Historical Experiments in Students' Hands: Unfragmenting Science through Action and History", *Science and Education* 17, 7 (2008), pp. 717–749.

5 Eleanor Duckworth, *"The having of wonderful ideas" and other essays on teaching and learning*, Teacher's College Press, New York, 1986/2006. For curricular materials, scholarly writing,

in the researches of Jean Piaget[6] and Bärbel Inhelder.[7] My observing, documenting, developing, and reflecting occurs alongside those of the students, and informs my thoughts for our further engagement with materials and questions. As an active participant in the evolving experience along with the students, I encourage the class as a space where we all:
- Explore what is around us;
- Consider and try out how others engaged with these things before us;
- Initiate experiments and understandings collaboratively with each other.

Historical scientific instruments, replicas and historical practices facilitate commencing experience where we act with something, find something unexpected, and create a further response. Through having historical instruments, or representations, in their hands, students frame relationships between their minds and bodies and things of the world in ways that are new for them. For example, measuring with a quadrant involves aligning one's eye with the quadrant and the distant object while a classmate notes where – within its arc – a weighted string hangs. Students' direct experience is inseparable from this measurement, unlike that of using many contemporary measurement devices.

One episode from a first day of a class illustrates how instruments from another time provide an entry to exploration. Often, on the first day of my class, we go outdoors and observe spatial relationships, sound and light. The weather was cold and clear and there was an auspicious circumstance: the class date coincided with a twice annual alignment of the setting Sun along the length of our building's hallway,[8] timed to occur in the final moments of our scheduled class. With this in mind, I chose to start with an activity that would lead us close to the sunset hallway. Being involved that day in reorganising archaic apparatus kept on back shelves from a previous era, one of my colleagues shared with me a strobe apparatus used by Harold Edgerton and originally designed for Albert A. Michelson's speed of light experiment.[9] On the spot, I decided to accept his offer to borrow it for class, along with two

---

and exploratory teachers' community, see https://cepress.org/ and https://criticalexplorers.org/ (accessed 13 Oct. 2020).

6 Jean Piaget, *The child's conception of the world*, J. & A. Tomlinson, trans., Littlefield, Adams, Totowa, NJ, 1960/1926.

7 B. Inhelder, H. Sinclair and M. Bovet, *Learning and the development of cognition* (trans. by S. Wedgwood), Harvard University Press, Cambridge, MA, 1974.

8 MIT Infinite Corridor Astronomy – MIT Henge, http://web.mit.edu/planning/www/mithenge.html (accessed 13 Oct. 2020).

9 Sperry Gyroscope Company produced the apparatus for Michelson's speed of light experiments; this one came to Harold Edgerton. Thomas P. Hughes, "Science and the Instrument-maker: Michelson, Sperry, and the Speed of Light", *Smithsonian Studies in History and Technology* 37 (1976), pp. 1–18.

SCIENTIFIC INSTRUMENTS IN EXPLORATORY TEACHING AND LEARNING   161

other old instruments from his collection. To these, I added a replica brass astrolabe,[10] an instrument that would be a continuing focus that term in the course (see *Section 5*).

After our course introductions, I laid these four unidentified instruments on the classroom table. Questions and wonder abounded: "It spins!" "Does light shine through!?" "Have you ever seen anything like that plug? Not me." "Can we take it apart?" "They didn't have superglue back then. How is it connected?"[11]

Encouraged by me, the students' curiosity moved their actions: turning the dials of the astrolabe replica (Fig. 9.1, top left); disassembly of the vacuum tube housing (Fig. 9.1, top right); shining a cell-phone light on the handgrip strobe; handling the revolving Sperry mirror unit (Fig. 9.1, bottom left); shining a cell-phone light at its revolving mirrors while seeking the reflection on a paper (Fig. 9.1, bottom right). Unfamiliar with these objects and their classmates, they became a team. Conjecturing, acting, and sharing, their investigations exemplify our course experiences. Their actions – taking initiative that may be more characteristic of nineteenth century science education[12] than subsequently – allowed the students to take control of their learning in a situation where most contemporary learners are expected to accept authoritative claims.

Through direct access to old instruments, as in this episode and in the examples below, the students related to objects present and past. For example, the act of sighting through a surveyor's level helped one student to realise its inadequacy for viewing indoors and to consider its historical use: "It was designed for large outdoor spaces [...] I wonder how they chose a starting location?"[13] By handling an instrument, students synthesised experience and meaning in ways that go beyond what they inferred from class readings and videos.

As experiences with instruments are provocative for my students, I arrange for visits with historical scientific instruments in various settings: exhibits at university and local museums; study or store-rooms at local museums and

---

10   Workshop of Norman Greene, Berkeley, CA; http://puzzlering.net/astrolabe.html (accessed 13 Oct. 2020).

11   Quotes from "EC.090 Transcript", MIT, February 6, 2018, manuscript: MIT, Cambridge, MA, Author's Archive. EC.050 (undergraduate level) and EC.090 (graduate level) are the Edgerton Center course numbers for "Recreate Historical Experiments, Inform the Future from the Past".

12   MIT's early laboratory text emphasises student agency: Edward C. Pickering, *Elements of Physical Manipulations*, Hurd and Houghton, New York, 1873, p. vi. See also: Peter Heering and Roland Wittje (eds.), *Learning by Doing: Experiments and Instruments in the History of Science Teaching*, Franz Steiner Verlag, Stuttgart, 2011, pp. 152, 189, 193–195 and 258–261.

13   Brian McCarthy, "Journal", EC.050, MIT, October 4, 2011, manuscript: MIT, Cambridge, MA, Author's Archive.

FIGURE 9.1 *From top left to bottom right*: Students explore an astrolabe replica (at right); disassemble a vacuum tube housing; examine the Sperry revolving mirror for Albert A. Michelson's speed of light experiments; and catch its reflections on paper

libraries; university laboratories and historical buildings; and at a collector's home, club, or business.

The MIT Museum hosts my class in a continuing relationship. Each term, MIT Museum curator Deborah Douglas and I brainstorm ideas for a class visit that involves students with the instruments, materials, culture and technology of the past. We draw on specific holdings or exhibits of the museum, as these relate to ongoing student activities. Class visits may occur in a gallery, a study room, an off-site storage depot – or move among all three! With Douglas's guidance, students may touch, hold, and use instruments formerly used in teaching and research at MIT. Often Douglas puts students in a detective role, even in investigating the function of instruments of which she has incomplete information. She asks students to observe materials, examine markings, operate a handle, follow a connection, read articles, speculate about possible functionalities. Along the way, she tells fascinating stories about instruments, interweaving insights about materials, politics, people, and discoveries – including her own.

SCIENTIFIC INSTRUMENTS IN EXPLORATORY TEACHING AND LEARNING 163

The following sections relate details of four principal ways that my students have engaged with historical scientific instruments:

- Viewing authentic instruments exhibited at the MIT Museum; responding, and even replicating.
- Interacting with authentic historical instruments in the MIT Museum storeroom.
- Using a model instrument in class; interacting with the original in a collection.
- Initiating major personal projects inspired by historical instruments.

2 Museum Gallery Visit and Responses

When the MIT Museum exhibits relate to course themes, my class visits the gallery. Gallery visits accommodate learners' participation with historical instruments by other modes than direct handling. Activities that are supported in galleries include: drawing and making diagrams; comparing instruments; photography; handling an analogue while viewing originals; discussion and questioning. In sessions following gallery visits, I invite students to share what piqued their interest and consider ways of continuing their curiosity in class.

For example, historical instruments were featured in the MIT Museum's 2006–2007 exhibit *Singular Beauty*, the first comprehensive public presentation of simple microscopes.[14] Drawn from Ray Giordano's collection, the instruments dated from the seventeenth to the nineteenth centuries; some elegantly crafted from precious materials; some by well-known optical instrument makers; others by unknown artisans. Two of my classes visited this gallery. One student, Mingwei Gu, participated in the first visit. What emerged through his interest informed our teaching with the next class of eighteen students.

Douglas led Gu and myself on a stroll through the gallery, marvelling at craft and designs (Fig. 9.2, top left). Douglas next directed us to one that did not stand out: a Leeuwenhoek copy.[15] Douglas amazed us with the wonders that Antoni van Leeuwenhoek saw with his similar instruments. Knowing of our class activities in remaking historical effects with everyday materials, she suggested we make a bead lens.

Intrigued, we took up the project. The extended process of finding suitable glass, heating and shaping it into a sphere, and observing through the resulting

---

14 Raymond Giordano, *Singular Beauty: Simple Microscopes from the Giordano Collection* (Catalogue of an exhibit at the MIT Museum, September 1, 2006–June 30, 2007), MIT Museum, Cambridge, MA, 2006.

15 Leeuwenhoek-type simple microscope, *Ibid.*, p. 13.

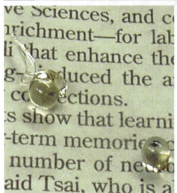

FIGURE 9.2 *From top left to bottom right*: Microscope gallery visit at the MIT Museum; Mingwei Gu blow-piping with Bunsen burner; his glass spheres; blow-piping by participants of the *XXVI Scientific Instrument Symposium*, Cambridge, MA, USA, 6–11 September 2007

bead immersed us in genuine exploratory work. With candle, alcohol lamp and Bunsen burner (Fig. 9.2, top right) as heat sources for blow-piping glass rods, Gu developed lung capacity and skill. Peter Houk, MIT's glass lab director, advised us on blowpipe technique. He made for us rods of soda lime glass, workable at a lower temperature than the Pyrex we tried first. Gu produced several glass sphere magnifiers (Fig. 9.2, bottom left). He demonstrated his expertise at a bead-making activity held at the 2007 SIC meeting.[16] The participants steadied their breathing, shaping flame and glass together, yielding a few glass spheres (Fig. 9.2, bottom right).

---

16   MIT Lab Tours, held on September 8, 2007, *XXVI Scientific Instrument Symposium*, Cambridge, MA, USA, 6–11 September 2007.

SCIENTIFIC INSTRUMENTS IN EXPLORATORY TEACHING AND LEARNING 165

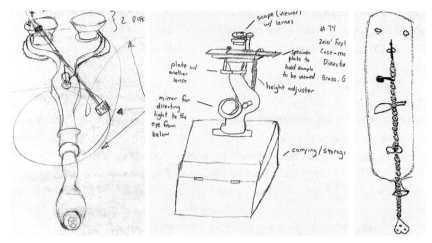

FIGURE 9.3 Three microscope drawings by (*from left to right*) Gabriela Antunes, Sharon Kiley and Lillian Rodriguez (October 5, 2007)

Months later, during the gallery visit of a class that I guest-taught at another school, Douglas invited students to draw on paper an image of a microscope that interested them. A display of all the drawings revealed diversity among the artefacts and in the observers' perceptions (Fig. 9.3). Students became invested in the instrument that they drew. One wrote: "I felt in love with one that stroked my imagination,[17] [...] questions came into my mind [...]. What is the story behind it? Was it used by common people?"[18] Feeling "confused" over how to sight through a compass microscope moved one student to ask: "are the [historical] users careful and detail oriented people?"[19] Another student advocated: "I hope women as well!" were among the makers of microscopes.[20] Impressed by a rock crystal elegantly mounted like a gem,[21] her classmate

17   The student refers to Hartsoeker's screw-barrel microscope; see Giordano, *Op. cit.* (n. 14), p. 14.
18   Gerald Koffi, "Microscope Writing Assignment", Honors Science 290, University of Massachusetts Boston, October 5, 2007, manuscript: MIT, Cambridge, MA, Author's Archive. Honors Science 290 is the number of the course that I taught during two fall semesters, at the request of the Honors Program.
19   John Kerpan, "Journal", Honors Science 290, University of Massachusetts Boston, October 5, 2007, manuscript: MIT, Cambridge, MA, Author's Archive.
20   Carolina Gomez, "The Microscope Activity", Honors Science 290, University of Massachusetts Boston, October 12, 2007, manuscript: MIT, Cambridge, MA, Author's Archive.
21   See: Giordano, *Op. cit.* (n. 14), p. 62.

imagined the scenario of its wealthy owner "whipping this out of his pocket" to show off at a party.[22] Students injected themselves into the historical context of the instruments, imagining what they would do with the instruments if they were the original owners.

A week later, in the classroom where I was a visiting instructor, students responded to the microscope exhibit by following up with their personal interests. Some explored magnification by placing different shaped lenses over printed text. Analysing and making diagrams of the light's path in differing microscopes, one student discerned three "completely different designs to capture light." Intrigued, he asked: "what would the same object look like when viewed in different microscopes? Different? Similar?"[23] Drawing on the findings of Gu's explorations, I provided materials that some students used in experimenting with shaping glass rods in a Bunsen burner flame.

These gallery visits create space where the students can contemplate historical instruments through personal observation and drawing, and by hearing stories. Students' connectedness to microscopes – initiated in the gallery – was extended in classroom explorations in diverse ways, as suggested above. These personal experiences sparked students' sense of dialogue among past, present and future humans doing science – including themselves. That students live deep qualities associated with "nature of science" is conveyed in one student's reflection:

> The fact that [historical people] made these microscopes speaks volumes of their curiosity [...] to observe the small worlds that exist among us and yet escape our natural eyes [...]. The human curiosity of wanting to know what is beyond one's own backyard, across the ocean [...] or even outside of space is the same with that of wanting to see the smallest details. [...] [Microscopy] has tremendous impact in our lives! I wonder what the makers of these early microscopes would think of today's super advanced microscopes.[24]

---

22 Renata De Carvalho, "Journal", Honors Science 290, University of Massachusetts Boston, October 5, 2007, manuscript: MIT, Cambridge, MA, Author's Archive.

23 Noam Shabani, "Journal", Honors Science 290, University of Massachusetts Boston, October 11, 2007, manuscript: MIT, Cambridge, MA, Author's Archive.

24 Gomez, *cit.* (n. 20).

3 Hands-On Storeroom Activities

The MIT Museum's off-site storage accommodates experiences where students encounter historical instruments that are not behind glass. Just to enter the vast warehouse is amazing (Fig. 9.4, top left). Amid towering shelves, packed with artefacts, old research apparatus and sculptures, with a human-powered plane hung from the rafters, students identify with the fictional character Indiana Jones![25] Douglas invites them to wander and select something to explore together. Scaling a ladder, a student brought down what resembled a violin case, an instrument he plays. When opened, a classmate recognised a historical Chinese balance, and demonstrated how her parents used one (Fig. 9.4, top right). Gary Stilwell (see *Section 5*), an advanced study fellow, gathered us around a model of the sailing ship *Maltese Falcon* built in 2006. He had seen the actual square-sailed ship close-up while crewing on the tall ship sv *Tenacious*.[26] He asked: "What's unique about it?"[27] Looking closely, classmates pondered the differences between sailing in the past and on the rigging-free yacht. Drawn by aesthetics and personal experience, students engaged each other in questioning and in understanding historical instruments.

For the bulk of each more than two-hour visit, Douglas prepares hands-on activities around themes of mutual interest. On a worktable, she places historical instruments that are unknown to the students. Gloves are distributed. Tools are available to assist in opening or testing instruments. Partway in, she may provide manuals, news articles, photos, and other materials from the historical time. Students discuss with her and each other while investigating the instruments' materials, culture and uses.

One winter, our class visit to the MIT Museum occurred after students had developed considerable experience observing the night sky. Every clear evening, we went to the Charles River Bridge, watching the sunset, waiting for Venus to appear, noticing birds, waves, ice and other wonders. Along with viewing by the naked eye and a portable refracting telescope, students estimated these bodies' elevation with: a training model sextant, a quadrant that students improvised from a protractor, and a laser-cut Galileo compass (see *Section 4*). They related these practices to readings about African-American astronomer

---

25   *Raiders of the Lost Ark*, 1981, directed by Steven Spielberg.

26   https://sailtraininginternational.org/vessel/tenacious/ (accessed 13 Oct. 2020).

27   Quote from "EC.090 Transcript", MIT, April 6, 2018, manuscript: MIT, Cambridge, MA, Author's Archive, fol. 4.

FIGURE 9.4 Students visit the MIT Museum's off-site storage. *From top left to bottom right*: Museum storeroom; testing a historical Chinese balance; opening boxes of surveying instruments; adjusting an instrument

Benjamin Banneker[28] and Scotsman James Ferguson's eighteenth-century astronomy manual.[29]

At that term's end, we went to the MIT Museum's off-site storage. In the light of Banneker's role in surveying Washington DC, Douglas chose to share nineteenth-century instruments used in training MIT's civil engineers. When we arrived, eight wooden boxes lay closed on the table. Douglas encouraged the students to walk about, open the boxes, examine each, and select one for personal study (Fig. 9.4, bottom left). On first impression, the students

---

28   Silvio Bedini, *The Life of Benjamin Banneker*, Maryland Historical Society, Baltimore MD, 1999.
29   James Ferguson, *Astronomy Explained upon Sir Isaac Newton's Principles*, James Ferguson, London, 1756.

identified the instruments as being like telescopes; one was labeled for survey-ing. Speaking on his choice, Jais Brohinsky said: "I have the least idea of what it did". Raul Largaespada, an aspiring aero/astro engineer, said he was drawn to one that most "looked like a telescope".[30]

Wearing gloves, the students were encouraged to remove the instruments from the boxes and explore whatever they could by viewing and manipulat-ing (Fig. 9.4, bottom right). Douglas provided a printed sheet for writing infer-ences, such as about materials, function, and era. Students were engrossed with what was before them, trying out the range of available motions (Fig. 9.5, left), sighting through lenses, and writing notes. Half an hour into the session, Douglas passed out historical civil engineering texts and manuals. On finding these instruments depicted, the students applied the texts in handling the optical levels and theodolites. These pursuits absorbed them as Douglas told stories about how surveying advanced Western empires.

"Using the apparatus, how tall is that shelf?" Douglas's direct question was met with silence. Her follow-up questions elicited tension. Unlike the activity of exploring instruments which engaged everyone, these questions required correct answers. Teacher-training student Brohinsky reflected: "it made me not want to interact [...] the tension was where the potential for real learning lay in that session".[31]

A story eased mutual strain. Largaespada shared a legendary fraternity prank. MIT freshman Oliver Smoot's body was lain end-to-end across the Bridge's length. Each placement was marked in paint. The "smoot" is now a unit of length in Google convertor![32] A student exclaimed: a wall-sized photo-graph in the storeroom depicts the original 1958 event! While observing on the bridge, we often walked over the colourfully painted "smoot" marks without noticing them. Now those marks attracted interest. This discussion uncovered history's evidences, previously unseen around us.

Lacking Smoot, Douglas handed a historical steel tape to the group. Using it to measure the floor entailed unexpected set-backs. The tape kinked, curled and did not stay taut. Its divisions did not start at "o"; part had broken off long ago. After measuring a horizontal distance along the floor with the tape, each student took a turn at looking through the optical level to read the vertical

---

30   Quotes from "EC.050 Transcript", MIT, January 27, 2017, manuscript: MIT, Cambridge, MA, Author's Archive, fol. 1.

31   Jais Brohinsky, "Two very curious wonderings from MIT Museum warehouse", EC.090, MIT, January 30, 2017, manuscript: MIT, Cambridge, MA, Author's Archive.

32   Patrick Gillooly, "Smoot reflects on his measurement", *MIT News* (September 24, 2008), https://news.mit.edu/2008/smoot-tt0924 (accessed 13 Oct. 2020).

FIGURE 9.5 Students visit the MIT Museum's off-site storage. *Left*: Handling an instrument; *Right*: sighting with an optical level

height on a surveyor's target (Fig. 9.5, right). Collaborating made the group aware of the back-and-forth inherent to forming scientific understanding.

This museum storage activity, which in Largaespada's words "helped familiarize us with the surveying tools and techniques Banneker would have used while surveying the land that would become Washington DC", interleaved supportively with our ongoing class activities. Direct experiences with telescopes, sextants, and other instruments pervaded those activities, from the first session's sighting of Venus, to personal observing with portable refracting telescopes, to our tour at MIT's Wallace Astrophysical Observatory on a clear night.[33] In Raul's regard, "our extensive class use of telescopes was a great boon in our quest to understand what Galilei and Banneker saw and calculated".[34]

In hands-on activities in the museum storeroom, like those described here, personal experience engages historical experience through physical actions with instruments and interpretive discussion. The learning goes beyond working out how the instruments were formerly used. It encompasses fun and surprise; insights about how pedagogy shapes an experience and how differing perspectives coordinate in collaboration. For example, along with questioning the tension arising between a "holder of knowledge and petitioners", doing the museum surveying activity stirred Brohinsky to propose an activity where

---

33 http://web.mit.edu/wallace/ (accessed 13 Oct. 2020).
34 Raul Largaespada, "Across Generations: Curiosity as a Unifying Force in EC.050", EC.050, MIT, February 7, 2017, manuscript: MIT, Cambridge, MA, Author's Archive.

young people would use ratio relations, based on their own bodies, to measure heights and distances.[35] He developed this idea into a lesson that he later taught in a summer camp.

4 Galileo Compass: Model and Original

Themes of Galileo Galilei, his instruments and experiments recur in my seminar.[36] A typical science classroom may have the physical materials for interpreting some Galilean era instruments – pendulum, ramp, telescope. Introducing other historical instruments, such as the astrolabe and Galileo's geometrical compass, in a classroom calls for constructing analogues having specific features. Initially, I addressed this circumstance with cardboard models. I assembled geometrical compasses from templates at the Museo Galileo website;[37] James Morrison produced two card astrolabes for the class.[38] On seeing my students struggle with these flimsy devices, my Edgerton Center colleague Ed Moriarty organised a student team to convert the templates to CAD as input to a laser cutter. After multiple iterations, they produced several laser-cut models in birch wood, of the geometrical compass at twice the original scale, and one enlarged astrolabe (40 cm diameter).

In opening a session with Galileo's compass, I distribute to each student pair: Galileo's manual,[39] a laser-cut geometrical compass, 12 inches (ca. 30 cm) long dividers, pencils and large paper.

Like cellphones, the geometrical compass is multipurpose. Often the students' first use of this is estimating a star's angle, then dialling that value into our large astrolabe. Other opening exercises in Galileo's manual include: finding a tower's height (see Douglas's question above), dividing a line in equal parts, and constructing polygons.

---

35   Brohinsky, *Op. cit.* (n. 31).

36   Philip Morrison, my teacher, inspired my passion for Galileo's instruments. I assisted him as researcher for Philip Morrison, Phylis Morrison, *The Ring of Truth*, Random House, New York, 1987; and the TV documentary *The Ring of Truth*, part 1: "Looking", 1987, available at www.youtube.com/watch?v=bQ4Oz2Xk2Ws (accessed 13 Oct. 2020).

37   "How to Make Galileo's Compass", Museo Galileo: Institute and Museum of the History of Science, Florence, https://brunelleschi.imss.fi.it/esplora/compasso/dswmedia/risorse/ecostruire_compasso.pdf (accessed 13 Oct. 2020).

38   James Morrison, *The Personal Astrolabe*, James Morrison, Rehoboth Beach, DE, 2010.

39   Galileo Galilei, *Operations of the Geometric and Military Compass*, Stillman Drake (trans.), 1977; available at https://brunelleschi.imss.fi.it/esplora/compasso/dswmedia/risorse/eleoperazioni.pdf (accessed 13 Oct. 2020). Also see the website simulation videos.

FIGURE 9.6 Sighting (*left*) and calculating (*right*) with a laser-cut Galileo compass

Without first reading Galileo's method, three engineering students set out in snow to measure a Boston skyscraper's height. Having sighted its angle with the laser-cut compass from two spots and paced a few steps between, they derived their own equations for the building's height. "Plugging in" their values and calculating by cellphone yielded an absurd value: 14 feet! The group devoted several sessions to track down errors. Redoing the experiment (Fig. 9.6, left) brought their attention to selecting locations for a perceptible difference in the plumb line's fall along the divided arc. A plausible height resulted when they applied Galileo's scales, which by encoding the trigonometry of the students' (revised) equations, reduced the calculation to ratios. The wooden model proved more reliable than the cellphone: one student, YouYou Li, discovered a reproducible malfunction in her cellphone's calculator.

Indoors, now trusting Galileo's manual as a guide, the group drew geometrical shapes using scales on the model compass to compute the length of a side (Fig. 9.6, right). We found that our model lacked the reference marks that Galileo describes as the key to constructing a square having the area equal to that of a section of a circle.[40]

Spurred on by these students' extensive involvement with Galileo's geometrical compass[41] and the discovery of the template's omission, I arranged a class visit to see an original: the fine instrument that Galileo presented to the Duke of Mantua, now held at The Collection of Historical Scientific Instruments,

---

40 *Ibid.*, pp. 77–78. These marks were omitted on the Museo Galileo template. After my inquiry, the template is now corrected.

41 The team created an educational music video featuring the cellphone, sextant, and astrolabe, *Science Enhances Romance*, posted under Entries 2014 at http://sciex.mit.edu/videos-all/ (accessed 26 Oct. 2020).

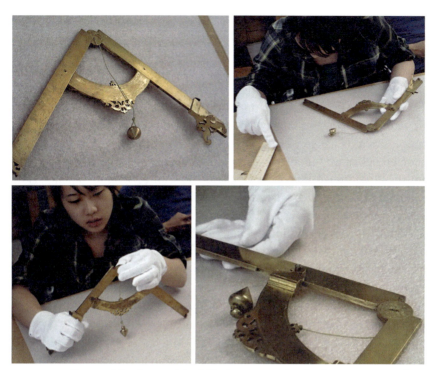

FIGURE 9.7   *From top left to bottom right*: Galileo's geometrical and military compass, Padua, c. 1604 (Harvard Collection of Historical Scientific Instruments, Cambridge, MA, inv. no. DW0950); comparing a laser-cut model with the original; orienting it for use as a quadrant; the flange covers the mark

Harvard University (Fig. 9.7, top left).[42] Being removed from display for our visit, curator Jean-François Gauvin put it in Li's gloved hands. She spied a fingerprint – was it Galileo's!? As mechanical engineering students, she and C.J. Munroe were impressed by how the compass arms' design conceals its construction, making it appear all one piece. In a crack that showed on one arm's side, not the other, they detected evidence of multipart makeup. Side-by-side, they compared the division markings on Galileo's instrument with those on the laser-cut model (Fig 9.7, top right). Discerning divisions, as Galileo's pupils

---

42   Galileo's geometrical and military compass, Padua, c. 1604, Harvard Collection of Historical Scientific Instruments, Boston, MA, inv. no. DW0950; http://waywiser.fas.harvard.edu/objects/3608/galileos-geometrical-and-military-compass?ctx=433cfbdd-853c-4c19-9b4c-fa5682e40921&idx=0 (accessed 13 Oct. 2020).

would need to do, is difficult: "its lines are so close [...] faint". Li exclaimed that one dent "really dug in!" likely made by a user's divider point.[43]

This instrument was configured as a quadrant, with an exquisite acorn-shaped weight suspended from its hinge. Its removable arc screwed onto the movable arms, fixing them at a right angle. In this configuration, the instrument cannot be used for calculation – as Li explained and demonstrated, with our model (Fig. 9.7, bottom left), to the curator who was unfamiliar with that aspect. The arc's flange was tightly screwed over where we expected to find the reference mark that was missing from our laser-cut model (Fig. 9.7, bottom right). Although that mark's position was unresolved by our hands-on examination, a new question arose as we considered the logistics of undoing the screw. How practical was it for a soldier in the field to sight with Galileo's instrument, then unscrew and calculate with it?

By exploring with the model compass outdoors and indoors, these engineers confronted failings in the modern tools and analysis they rely on, and gained respect for Galileo and skill with his instrument. Bringing this experience to the original instrument enabled them to comprehend the maker's design and discern marks, dents and attributes of how it was used. Personal experience with both model and original grounded the students' interactive questioning and vision of the instrument's role in historical hands.

5   Student-Initiated Projects Inspired by Historical Instruments

I encourage students to act on their curiosity in developing explorations. Some projects extend beyond my course. Here I summarise two such projects – one inspired by the other – where the students' passion for historical instruments creates new experiences for us all.

For architect Francesca Liuni, who entered my seminar intent on reinterpreting Islamic science, our class activities with astrolabes struck a chord of lasting resonance. While discussing SIC's 2015 theme "Instruments in Conflict" with classmate Ronald Heisser in relation to their poster proposals,[44] Liuni voiced her emergent vision for a radically interactive museum. It would be

---

43   Quotes from "EC.050 Transcript", MIT, April 1, 2014, manuscript: MIT, Cambridge, MA, Author's Archive, fol. 9.

44   Francesca Liuni, "Balancing the Astrolabe between Art and Science"; Francesca Liuni and Ronald Heisser, "Conflict and Balance: Archimedean Conflicts"; Ronald Heisser, "Methodical Conflict in Past and Present Fluid Mechanics Research: An Alternative Investigation"; posters presented at *XXXIV Scientific Instrument Symposium*, Turin, Italy, 7–11 September 2015.

SCIENTIFIC INSTRUMENTS IN EXPLORATORY TEACHING AND LEARNING   175

an immersion in the astrolabe: using it, being in its space, and with historical astrolabes displayed. When shared with her advisor, he reacted "you are crazy!"[45]

Liuni was undeterred. Her first model puts museum-goers amid windows incised with the astrolabe's grid projection of the celestial sphere's azimuthal lines and altitude circles, so as to cast their shadow arcs within the interior space.[46] SIC members' feedback fuelled her next version: an award-winning master's thesis uniting art and science in engaging the public with the astrolabe.[47]

Becoming immersed in the geometrical thinking of Islamic science, Liuni discovered its architectural potential. Responding to Walter Benjamin's vision,[48] where past and present meet in recreating historical experience, she converted the axiomatic system underlying the astrolabe into architectural structures through which today's viewers encounter its constraints and possibilities. Her intricate hand-crafted thesis models and architectural drawings place the viewer within a celestial sphere, bounded yet open, and wandering among geometrical relationships of angles in space (Fig. 9.8). Architecture, she reflected, integrates "the abstraction and intuitive language of Art with the exactitude and practicality of Science".[49]

Liuni's next project (Fig. 9.9) echoes her original dream. In a gallery at Harvard, she created and built a walk-in spatial rendering of the divided sphere.[50] Posters depicting her artistic renderings of the astrolabe's projections lined the walls, while an authentic seventeenth-century Persian astrolabe was exhibited within.

I invited Liuni to share her astrolabe work in my next year's seminar. Never having seen an astrolabe before, yet trained with the sextant, Gary Stilwell (see *Section 3*) scrutinised our laser-cut astrolabe (Fig. 9.10, above), eventually identifying the North Star. With breath-taking realisation, a classmate exclaimed: "when they invented the astrolabe, they thought the Sun moved and Earth does not! Wow!" Stilwell extended that perspective-taking to the entire solar system:

---

45   Quote from "EC.050 Transcript", MIT, March 12, 2015, manuscript: MIT, Cambridge, MA, Author's Archive, fol. 11.

46   Liuni, *Op. cit.* (n. 44).

47   F. Liuni, *Experiencing Mathematical Proves: Syntax of an Astrolabe*, Master's Thesis, MIT, Cambridge, MA, 2016; https://chsi.harvard.edu/files/chsi/files/967222067-mit.pdf (accessed 13 Oct. 2020).

48   Walter Benjamin, "Theses on the Philosophy of History", in Hannah Arendt (ed.), *Illuminations*, Fontana-Collins, [London], 1973, pp. 255–266.

49   Liuni, *Op. cit.* (n. 47), p. 51.

50   F. Liuni, *Syntax of an Astrolabe*, Foyer Gallery, Harvard, 2017; https://chsi.harvard.edu/syntax-astrolabe (accessed 13 Oct. 2020).

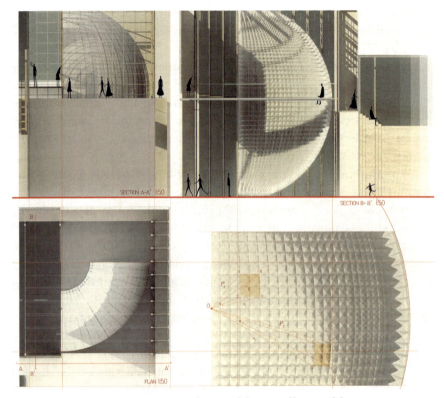

FIGURE 9.8   Francesca Liuni's section architectural drawing of her astrolabe museum
F. LIUNI, *EXPERIENCING MATHEMATICAL PROVES: SYNTAX OF AN ASTROLABE*, MASTER OF SCIENCE IN ARCHITECTURE STUDIES THESIS, MIT, CAMBRIDGE, MA, 2016, PP. 66–68

"What if we made eight astrolabes, one for each planet!?"[51] Stilwell developed understanding for acting on this idea by constructing his astrolabe geometry by hand on paper with guidance from Alistair Kwan's manual.[52] Our class was awed by Stilwell's term project. He coded, designed and printed on paper, an astrolabe tympanum for each planet in our solar system. The base location for each planetary tympanum corresponds to the latitude of Cambridge

---

51   Quotes from "EC.090 Transcript", MIT, February 15, 2018, manuscript: MIT, Cambridge, MA, Author's Archive, fol. 10.
52   Alistair Kwan, *Introduction to Astrolabe Construction and Application*, 2007, MS.; A. Kwan, "Determining Historical Practices through Critical Replication: A Classroom Trial" *Rittenhouse* 22, 2 (2008), pp. 132–151.

FIGURE 9.9 Francesca Liuni explains her exhibit, *Syntax of an Astrolabe*, Foyer Gallery, Harvard, Cambridge, MA, 2017

Massachusetts, on that planet (Fig. 9.10, below).[53] My subsequent students are inspired by viewing Stilwell's exhibits of his planetary astrolabes, printed in crisp lines on titanium, copper, glass and paper.[54]

The projects of Liuni and Stilwell were generated and shaped through their own questions: "How can I create an historical experience?" "What if we made eight astrolabes?" Personal curiosity shaped the iterative and evolving routes

---

53 Gary Stilwell, "Le Voyage Dans Les Planets", EC.090, MIT, May 2018, manuscript: MIT, Cambridge, MA, Author's Archive.
54 G. Stilwell, *Where is Here? Astrolabes for Future Use on Mercury, Venus, Mars, Jupiter … and even Pluto*, Wiesner Student Art Gallery, MIT, Cambridge, MA, December 2018; and ProjXpo MIT, Cambridge, MA, January 31, 2020.

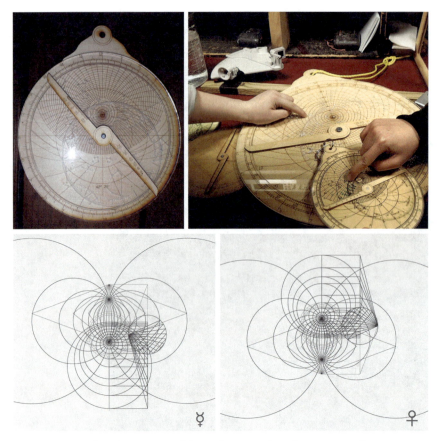

FIGURE 9.10 *Above*: Laser-cut astrolabe in class: incised acrylic rete, 40 cm diameter; *Below*: Gary Stilwell's line drawings with stereographic projections corresponding to astrolabe plates for the planets Mercury and Venus, for an observer at 40° north latitude on each planet

by which each student delved into the astrolabe's mathematics, geometry, design, history, and astronomy. Drawing by hand with paper and pencil, they sketched and constructed its geometries anew for themselves. Each transformed their vision of the astrolabe into digital media: Liuni by architectural and spatial renderings; Stilwell by *JavaScript* coding, taking into account each planet's orientation. Having traversed mathematical, constructive, abstract and digital realms, each project brings the viewer to a new physical experience with the astrolabe: Liuni's through architectural space; Stilwell's through lines inscribed on polished metals and glass, helping future travellers find their way in outer space with a historically inspired instrument.

6 Reflections on Historical Instruments in Exploratory Education

Thrill is palpable among learners engaging with authentic historical scientific instruments, as illustrated here. That depth in responsiveness arises in diverse forms of interaction: whether with an instrument on display ("I felt in love with one"), or held in a gloved hand ("Is that Galileo's fingerprint?"),[55] or in active use (sighting a surveyor's target through an optical level "I think I see the circle!").[56] Emotion, aesthetics, bemusement and action with historical instruments figure in how learners bring themselves into relation with the past, its people and science.

Where historical instruments are not museum quality, the opportunities widen for students to handle and trial uses, enhancing their participation with historical materials and questions, as in this chapter's first example. Models of historical instruments, such as the blowpipe and laser-cut Galileo compass and astrolabe, afford multiple and distinct interactions over time, as was needed for Mingwei Gu to develop blow-piping expertise, and for YouYou Li and her team to work through their uncertainties in observation methods and analysis, eventually finding a building's height with a model of Galileo's geometrical compass. Practical skill with the instrument, discovering all its functions and adapting to its limits, such as they gained, requires more than a single museum visit to attain. Through activities such as observing Venus from the Charles River Bridge and working hot glass, students meet first-hand the phenomena that historical instruments address.

Awe for the antiquity of human observing and awareness of taking part in it now, opens a dialogue among students and their predecessors. That relation invites students to put their own past and future into relief. Such perspective-taking is especially poignant where their predecessors were students like them: Galileo's pupils using the geometrical compass; MIT civil engineering students working with the same surveying instruments; and Oliver Smoot, marks of whose body measurements still remain on the Bridge where they walk now. Sharing her personal transformation arising through exploratory experiences with science and history, one student reflected: "This class opened my eyes, mind, and thoughts to explore things that I never thought I would. Now when I step, I see a lot more and now I question things, like Galileo did".[57]

---

55   "EC.050 Transcript", cit. (n. 45), fol. 6.

56   "EC.050 Transcript", cit. (n. 30), fol. 5.

57   Renata De Carvalho, "Reflection Paper", Honors Science 290, University of Massachusetts Boston, December 20, 2007, manuscript: MIT, Cambridge, MA, Author's Archive.

Each student's relation with the past, its science, instruments, and the relief it sets upon today's context, is stirred by experiences that put them in the role of detective, investigator, observer, or surveyor. In the sessions described here, education is inherent to experiences of exploring in the unknown among learners together with teachers and others of the past. Historically, instruments were made and used by those going into the unknown. Those same instruments can open to further unknowns today, where students' experience with them is an unknown by which they evolve and learn – about historical instruments and what matters in their present world. To the extent that students have and take agency to explore, question and act, the experiences they live and create will widen all our understanding.

Acknowledgements

Thanks to the Students: Jais Brohinsky, Gabriela Cordeiro-Antunes, Renata De Carvalho, T.J. Dimacali, Carolina Gomez, Mingwei Gu, Ronald Heisser, John Kerpan, Sharon Kiley, Gerald Koffi, Raul Largaespada, Ilia Lebedev, Tongji Li, Francesca Liuni, Brian McCarthy, C.J. Munroe, Yoonah Park, Lillian Rodriguez, Gary Stilwell, Summer Xia, Jinwen Ye, and others. Thanks for supporting class activities: MIT Edgerton Center, Jim Bales, Sandi Lipnoski, Ed Moriarty, Tony Caloggero; MIT Museum, Deborah (Debbie) Douglas; MIT Glass Lab, Peter Houk; Harvard Collection of Historical Scientific Instruments, Jean-François Gauvin; University of Massachusetts Boston Honors Program, Rajini Srikanth. Laser cut instruments were developed by Ed Moriarty, Adrian Tanner, Jackie Sly, Alban Cobi, Peter Moriarty, Charles Guan. For encouragement in investigatory teaching and learning, I thank Peter Heering, Eleanor Duckworth and Alva Couch. Memory: Philip Morrison, Alanna Connors, Yan Yang, my parents.

CHAPTER 10

"What Is Happening in the Lab?" Transforming the School Laboratory into a Contextual Science Teaching Environment

Flora Paparou | ORCID: 0000-0001-7733-8198

Introduction

What is happening in the lab? was the title of the science theatre performance which I co-ordinated in the academic year 2019–2020 at the 1st Upper High School (Lyceum) of Vrilissia, Athens.[1] A group of twelve school students from the first and second grades (15 to 17 years old) participated in its creation. In the different scenes of the play, the audience was invited to participate in a 1660s Royal Society meeting, to attend an eighteenth-century public demonstration on electricity, and to follow Jules Verne's visit to Ruhmkorff's laboratory in nineteenth-century Paris. Strange explosions, which erupted between these differently set scenes brought the inimitable investigator, Sherlock Holmes, to the stage to identify "what's happening in the lab". The play, which was prepared inside the science laboratory, was to be performed to an audience of students and teachers from a French school who were scheduled to visit ours in March 2020, but all such activities were cancelled due to the Covid-19 pandemic. When the school closed, our rehearsals were at the final stage. As the participation of the audience was crucial for the unfolding of the show, students and teachers from our school were invited to the rehearsals to help the actors imagine what the French visitors' reactions might be.

Some years ago, a similar scenario was used for the creation of an escape-room game, also addressed to high-school level audiences, which was presented at the *Athens Science Festival 2016*. It was a science communication project, the result of the collaboration of our school with the 4th Lyceum of Palaio Faliro, Athens.[2] The attenders to the Science Fair escape-room game were invited to

---

1　This chapter is based on the identically-named presentation at the *XXXIX Scientific Instrument Symposium*, London, UK, 14–18 September, 2020. Part of its content was included in Flora Paparou, "Lecture Demonstrations in History of Science Museums", *XXXIV Scientific Instrument Symposium*, Turin, Italy, 7–11 September 2015.

2　The escape-room project was conducted during the academic year 2015–2016. A group of twenty-five students aged from 15 to 18 years old (eleven students from our school and

---

© FLORA PAPAROU, 2022 | DOI:10.1163/9789004499676_012

explore a series of connecting rooms that were supplied with equipment to represent places where experiments were carried out during the seventeenth and nineteenth centuries. The game involved touring such historical rooms as: Galileo's study, Boyle's workspace (Fig. 10.1, top) and Newton's rooms, a natural magician's shop, Magdeburg open spaces during the performance of the famous eponymous experiment (Fig. 10.1, bottom),[3] and the Royal Institution of Great Britain in the late nineteenth century. A sequence of puzzles associated with the experiments, the scientists, the instruments, the laboratories, and the dates comprised the navigation throughout the whole escape room, which took the form of an interactive museum. In some rooms the participants had to accomplish the experiments themselves. In others they became active audiences to theatrical events, which imitated historical experimental demonstrations. In the Royal Institution room, students from our school enacted John Tyndall and Lord Rayleigh delivering a lecture demonstration on the colours of the sky, the inert gases, and spectroscopy (Fig. 10.7, above). While the visitors were carefully attending to the performers, a woman planted in the audience fell suddenly to the ground. Sherlock Holmes, who was also present, spread newspaper information on a strange lethal disease afflicting London. He presented a nineteenth-century London map noting where the victims were found and an emission spectrum which was the result of spectroscopic analysis of a victim's blood. The audience had to combine what Lord Rayleigh had just presented on spectroscopy with the information provided by Sherlock Holmes to identify the cause of the disease: the culprit was mercury. Traces of the metal having escaped for centuries from the scientists' laboratories had contaminated the London water system.

Both scenarios sketched above intertwine issues from the history of science, the history of experimentation, and the nature of scientific instruments. In the course of their development, the instruments of the school laboratory became magical kaleidoscopes, through which historical scientific instruments and experiments were focused and explored. The purpose of this paper is to

---

fourteen from the 4th Lyceum of Palaio Faliro) participated in it. Our school group was coordinated by me (science teacher) and Katerina Balomenou (art teacher). The collaborating school group was coordinated by Vasiliki Milioni (science teacher) and Polyxeni Simou (Greek language and theatre teacher).

3   "The Experiment Demonstrating that as a Result of Air Pressure, Two Hemispheres can be Joined Together in Such a Way That They Cannot be Separated by Sixteen Horses", in: Otto von Guericke, *The New (So-Called) Magdeburg Experiments of Otto von Guericke*, « Archives Internationales D'Histoire des Idées / International Archives of the History of Ideas 137 », Springer Verlag, Dordrecht, 1994, pp. 160–162.

FIGURE 10.1   Athens Science Festival 2016. *Top*: Re-enactment of a Royal Society meeting of the 1660s; *Bottom*: The Magdeburg experiment

explain how activities focusing on the material culture of science, on science communication and on the contextual approach of scientific topics, created a new tradition for the teaching of experiment in our school and rendered the science laboratory a pole of attraction for students of different interests and inclinations.

2 Theoretical Challenges, Practical Problems and Solutions Chosen

In response to the finding that a low percentage of students held positive attitudes towards science, educators of the western world in the 1980s proposed interdisciplinary science curricula. The role of the history and philosophy of science has been upgraded in these reformed curricula, and those two subjects functioned as design axes for the creation of contextual science teaching proposals.[4] Reflecting the stance of history and philosophy at that time, these curricular proposals implemented those fields by focusing mainly on matters relating to changes in the theories and concepts of science. Even if historical case studies of science experiments were introduced in the classroom, or in the academic literature on which the classroom curricula were based, the role of instruments in those experiments, and philosophical perspectives on experiments were neglected. This educational trend changed in the 1990s, due to the impact of 1980s research in the philosophy of experiment.[5] A new historiography of science emerged, where experiment has its own life. Experiment came to be regarded as a complicated process, having technical, theoretical, and social components. As Peter Galison explains:

> We need a historiography with room for a multiplicity of cultures [...]: a culture of theory, surely, but also a culture of experimentation, and a culture of instrument building [...]. We need a history of experimentation that accords that activity the same depth of structure, quirks, breaks, continuities, and traditions that we have come to expect from theory.[6]

This new historiography concurrently transformed both academic research and educational practices. Researchers came to value and examine the material culture of science. Educators came to value and undertake the teaching of experiment. Influenced by these new perspectives, innovative educational projects in museums, universities and schools were initiated to: reconstruct historical scientific instruments, replicate historical experiments, and conduct science communication events for the general public.

---

4 See: Michael R. Matthews, *Science Teaching: The Role of History and Philosophy of Science*, Routledge, New York and London, 1994, pp. 35–48.

5 See, for example, Ian Hacking, *Representing and Intervening, Introductory Topics in the Philosophy of Natural Science*, Cambridge University Press, Cambridge, UK, 1983.

6 Peter Galison, "History, Philosophy and the Central Metaphor", *Science in Context* 2, 1 (1988), pp. 197–212: p. 211.

However, despite these contributions, in today's Greek General Lyceum,[7] the teaching of experiment is confined to a small percentage of the applied science curriculum. The main tradition for the introduction of experiments in typical science courses is utilitarian. The function of the experiment that the students are expected to perform is limited to an exercise that confirms an already explained theory.

From my position as a science teacher, I sought to alter this prevailing tradition of downgrading the role of experiment in the teaching of science. Firstly, I systematically used experimental demonstrations and laboratory practice as a means to spark students' interest in topics that are included within the framework of my typical science courses. Secondly, I launched experimental enquiries within the framework of "Research Creative Activities", a course that was part of the national Greek curriculum until 2019–2020 and required that the teacher design its content and activities in accordance with the students' interests.[8] Thirdly, I organised extracurricular laboratory projects.[9] Little by little, the message that "something is happening in the lab" spread across the community of my school. Students with an interest in science realised that there was a way to deepen their scientific knowledge, to cultivate experimental skills, and to work in group settings towards participating in science competitions or science communication events. Since 2012, more than 150 students have joined these curricular and extracurricular laboratory projects.

3 Creating a New Tradition for the Teaching of Experiment in the School Laboratory

History and philosophy of science function as design axes in our projects, according to the examples of the 1980s curricular reforms for contextual science teaching. Moreover, being inspired by the 1990s focus on the material culture of science, our initiatives entail an additional perspective, one that

---

7 Lyceum is the upper secondary education school. Its students are aged from 16 to 18 years old.
8 The "Research Creative Activities" course was introduced in the Lyceum curriculum in 2011, and was addressed to students of the first two grades, that is 15–16 and 16–17 years old. The course was initially applied for two hours a week. From 2013–2014 until 2019–2020, it was reduced to one hour per week for the second grade.
9 The extracurricular courses are voluntary for students and teachers in Greek schools. If the teachers choose to organise such a course, they have to create a two-hour programme per week, after school hours. The 2015–2016 and 2019–2020 projects mentioned in the introduction were conducted within the framework of extracurricular laboratory programmes.

changes the way the laboratory instruments are usually perceived. Besides their practical use, instruments figure in our activities as representatives of historical objects in books or museum galleries. These objects possess biographies of their own, which link historical facts, persons, places and events. For example, our classroom vacuum pump, an instrument of the 1970s, was presented to my students within the context of the seventeenth-century research on air, vacuum and atmospheric pressure. Around this object, students and I wove associations from seventeenth-century engravings, virtual tours in history of science museums, historical scientific documents and experiments. Similarly, we engaged with our classroom Wimshurst electrostatic generator while drawing on the history of electrostatic generators and eighteenth- and nineteenth-century electricity shows. Analogously, our Ruhmkorff coil transported us to the nineteenth-century scientific stage. We interacted with it by way of its role in the impressive experiments on electrical discharge in rarefied gases, the controversies on the nature of the cathode rays, and the development of spectroscopy as an essential analytical tool.

To realise our laboratory projects the re-classification and improvement of the laboratory equipment at our school was required. Besides their usual classification under the categories of mechanics, fluid dynamics, thermodynamics, static and dynamic electricity, electromagnetism, optics, chemistry, and others, the instruments had to be ranked according to their degree of resemblance to historical scientific equipment. This analysis revealed where the array of our laboratory equipment was insufficient for the goal of allowing students to conduct multiple series of experiments relating to historical scientific researches on air, vacuum, electricity and astronomy. New instruments were purchased, and simple devices were reconstructed, some of them with the help of the students.[10] Thus, students got the opportunity to study and perform experiments from diverse historical and physical contexts. Phenomena of vacuum and atmospheric pressure were considered and demonstrated in the

---

10   Among the instruments reconstructed were Nollet's double cone, Leyden jars, an early electricity electric-bells apparatus, a thunder-house, a Volta pile, and the inside part of Kepler's and Galileo's telescopes. Some of these devices, such as a Leyden jar made of glass and metal and the thunder house, were reconstructed in collaboration with technicians. Others, such as the Volta pile and the inside of the telescopes, were created by me, and others, such as Leyden jars with modern materials, the Franklin's bells apparatus, and Nollet's double-cone, were constructed with the help of the students within the framework of curricular and extracurricular projects.

"WHAT IS HAPPENING IN THE LAB?" 187

context of researches by Gasparo Berti,[11] Otto von Guericke,[12] the Accademia del Cimento,[13] and Robert Boyle.[14] The students were guided to conduct experiments inspired by Galileo's writings on falling bodies,[15] and by Newton's papers on the analysis and synthesis of light.[16] They were introduced to the performance of experiments on static electricity based on the works of Jean Antoine Nollet, Pieter van Musschenbroek, Benjamin Franklin, Alessandro Volta and eighteenth- and nineteenth-century public lecturers.[17] Students were invited to explore the history of the electron, with the help of demonstrations based

11 See: William R. Shea, *Designing Experiments and Games of Chance: The Unconventional Science of Blaise Pascal*, Science History Publications, Canton, 2003, pp. 24–31; William Edgar Knowles Middleton, *The History of the Barometer*, The Johns Hopkins Press, Baltimore, 1964, pp. 10–18; W.E.K. Middleton, "The Place of Torricelli in the History of the Barometer", *Isis* 54, 1 (1963), pp. 11–28.

12 O. von Guericke, *Op. cit.* (n. 3), pp. 160–165.

13 Lorenzo Magalotti, *Essayes of natural experiments: made in the Academie del Cimento, under protection of the most serene Prince Leopold of Tuscany written in Italian by the secretary of that academy, Englished by Richard Waller* (facsimile of the 1684 ed.), «The Sources of Science», Johnson Reprint Corporation, New York and London, 1964.

14 James Bryant Conant, "Robert Boyle's Experiments in Pneumatics", in J.B. Conant and L.K. Nash, *Harvard Case Histories in Experimental Science*, Harvard University Press, Cambridge (MA),1957–1966, 2 vols.: v. 1, pp.1–63. See also, Robert Boyle, *The philosophical works of the Honourable Robert Boyle, esq., abridged, methodized and disposed under the general heads of physics, statics, pneumatics, natural history, chymistry and medicine by Peter Shaw, M.D.*, London, Printed for W. and J. Innys and J. Osborn and T. Longman, 1725, 3 vols.: v. 2, pp. 405–524.

15 Galileo Galilei, *Two New Sciences. Including Centers of Gravity and Force of Percussion* (trans. by Stillman Drake), Wall & Emerson Inc., Toronto, 1989[2], pp. 65–79.

16 Isaac Newton, "A Letter of Mr. Isaac Newton, Professor of the Mathematicks in the University of Cambridge; containing his New Theory about Light and Colors", *Philosophical Transactions of the Royal Society* 80 (19 Feb. 1671/1672), pp. 3075–3087; Isaac Newton, *Opticks*, Dover Publications Inc., New York, 1979, pp. 154–158.

17 For eighteenth-century electricity experiments see: Duane Roller and Duane H.D. Roller, "The Development of the Concept of Electric Charge", in J.B. Conant and L.K. Nash (eds.), *Op. cit.* (n. 14), v. 2, pp. 543–608. For the other authors mentioned above see: Jean Antoine Nollet: "Observations sur quelques nouveaux phénomènes d'Electricité", *Mémoires de l'Académie Royale des Sciences*, Paris, 1746, pp. 2–23; Benjamin Franklin, "From Benjamin Franklin to Peter Collinson, September 1753", *Founders Online*, National Archives, https://founders.archives.gov/documents/Franklin/01-05-02-0021 (accessed 7 Nov. 2020); Leonard W. Labaree (ed.), *The Papers of Benjamin Franklin*, Yale University Press, New Haven, 1959–1969, 13 vols.: v. 5 (1962), "July 1, 1753, through March 31, 1755", pp. 68–79; Alessandro Volta and Joseph Banks, "On the electricity excited by the mere contact of conducting substances of different kinds", *Philosophical Magazine* 7, 28 (1800), pp. 289–311.

on Crookes' lecture on radiant matter,[18] and Thomson's crucial experiment on the nature of cathode rays.[19] Also, Kirchhoff and Bunsen's 1860 paper was used to approach spectroscopy experiments,[20] and Tyndall's demonstration to simulate the blue of the sky.[21] The students were given short excerpts from the texts cited above translated into Greek, or summaries of the texts.[22] In many cases, among which are those of von Guericke's and Berti's experiments, the students were asked to decode the information included in the details of historical illustrations and engravings.[23]

Within the framework of our laboratory projects, issues of the nature of scientific instruments were examined. The students were guided to distinguish between different kinds of instruments. The vacuum pump, the electrostatic generators and the Ruhmkorff coil were highlighted, not only because they had central roles in long-time and fruitful experimental researches,[24] but also as representatives of the Baconian tradition. From that perspective, these are instruments which enable the generation of phenomena. Such instruments as the thermometer and the barometer were presented to the students as representatives of the measuring tradition, whereas instruments, such as the lenses and the prisms, were highlighted for their multiple roles.[25] The students had

---

18   William Crookes, *On Radiant Matter: A lecture delivered to the British Association for the Advancement of Science, at Sheffield, Friday, August 22, 1879*, E.J. Davey, London, 1879.

19   Joseph John Thomson, "Cathode Rays", *Philosophical Magazine* 44, 293 (1897), pp. 293–316.

20   Gustav Kirchhoff and Robert Bunsen, "Chemical Analysis by Observation of Spectra", *Annalen der Physik und der Chemie* 110 (1860), pp. 161–189.

21   John Tyndall, "The Blue of the Sky", in: J. Tyndall, *Six Lectures on Light. Delivered in the United States in 1872–1873*, Longmans & Co., London – New York – Bombay, 1906, pp. 149–152.

22   The excerpts of the translated texts and the summaries given to the students were short, *i.e.* of one or two pages at maximum. Most of the translations were done by me. See: Flora Paparou, *The Utilisation of the History and Philosophy of Science in Science Teaching – Developing Educational Activities around Historical Scientific Instruments*, PhD Thesis, National and Kapodistrian University of Athens, Academic year 2012, pp. 384–413 (in Greek), available at: https://thesis.ekt.gr/thesisBookReader/id/35259#page/6/mode/2up (accessed 7 Nov. 2020).

23   For Berti's experiment, see: Middleton, *The History of the Barometer, cit.* (n. 11), p. 11; Shea, *Op. cit.* (n. 11), pp. 25–27. For the Magdeburg experiment see: von Guericke, *Op. cit.* (n. 3), p. 162, pl. XI, and p. 165, pl. XII.

24   The vacuum pump had a prominent role in the seventeenth-century research on air and atmospheric pressure. The electrostatic generators and the Ruhmkorff coil had prominent roles, respectively, in the eighteenth- and nineteenth-century researches on electricity.

25   Being tools of natural magic, in Newton's hands the prisms became fundamental instruments to produce matters of fact; see: Simon Schaffer, "Glass works: Newton's prisms and the uses of experiment", in D. Gooding, T. Pinch and S. Schaffer (eds.), *The Uses of Experiment*, Cambridge University Press, Cambridge, UK, 1989, pp.67–104. During the

"WHAT IS HAPPENING IN THE LAB?"					189

the opportunity to explore different experimental roles and traditions. They worked on both *trials*, involving investigations into phenomena and behaviours that were unknown to the experimenter, and *demonstrations*, where the outcome of the experiment is known in advance by those who perform it.[26] They practised experiments having a role in inquiry, testing or measuring.[27]

Our projects highlighted the social context of scientific developments. Among the issues that were extensively explored were: the crucial role of witnessing for the establishment of experimental science in the seventeenth century;[28] public demonstrations as a means of science-dissemination in the eighteenth and nineteenth centuries;[29] nineteenth-century "scientific novels" and writings – like those of Jules Verne and Herbert George Wells – as indications of the significant cultural impact of science in society; the twentieth century as an era when many utopias and dystopias about science in society took shape and form.[30]

---

    nineteenth century, the prism became the core of the spectroscope, a basic precision instrument. The lenses had also multiple roles, being used as tools aimed at either deceiving or expanding the senses.

26   Historical trials and demonstrations represent different experimental cultures: scientific investigation and science communication. Few experts were engaged in the performance and witnessing of trials, when the scale of experimentation was still small. On the contrary, the demonstrations were science communication events, and were performed in front of larger audiences, either of experts or lay public. For seventeenth-century trials and demonstrations, see: Steven Shapin, "The House of Experiment in Seventeenth-Century England", *Isis* 79, 3 (1988), pp. 373–404.

27   With the help of enquiry experiments the students approached the exploratory experimental tradition, in which the creation of phenomena has a prominent role. With the help of testing experiments the students approached the classical experimental tradition, in which the experimental process is aimed at confirming an already expressed theory. Finally, they approached the measuring experimental tradition by using precision instruments or by conducting experiments aimed at measuring physical magnitudes. For the different experimental traditions see: Friedrich Steinle, "Experiments in history and philosophy of science", *Perspectives on Science* 10, 4 (2002), pp. 408–432. See also: Theodore Arabatzis, "Experiment", in Maryanne Cline Horowitz (ed.), *New Dictionary of the History of Ideas*, Charles Scribner's Sons, New York, 2005, 6 vols.: v. 2, pp. 765–769.

28   See: S. Shapin, "Pump and Circumstance: Robert Boyle's Literary Technology", *Social Studies of Science* 14, 4 (1984), pp. 481–520.

29   See: Bernadette Bensaude-Vincent and Christine Blondel (eds.), *Science and Spectacle in the European Enlightenment*, Aldershot, Ashgate, 2008.

30   The literature cited above, concerning the social components of scientific developments, was not given to the students in the form of extended texts. It constituted my theoretical framework and was communicated to the students through lecture demonstrations, short excerpts from texts translated into Greek, movies and discussions. The students did incorporate aspects of this framework in their experimental projects and science communication scenarios.

4 Teaching Methodology and Case-Studies

Although our various laboratory projects had similar design axes, each year the specific activities were different as they depended on the students' interests. I will describe some of the most impressive results of our projects, as well as their common backbones. Each year at the beginning of both curricular (the "Research Creative Activities" course) and extracurricular laboratory courses, our laboratory was transformed into an exhibition room. In it, a series of experiments was presented contextually through lecture demonstrations. To deepen their knowledge on the experiments presented, students were given further information, such as a bibliography, audio-visual material, historical texts and engravings. In this preliminary phase, the instruments of our laboratory were linked to instruments of history of science museums, which in many cases were to be visited during the ensuing projects. Then, the second phase began. The students were asked to form groups of three to five people. They could choose their partners and the topics to be explored. The students learned the operation of the machines included in their projects and redid, with the help of these machines, the experiments presented in the introductory lectures. Then they were free to change and improve the experimental set-ups. They were also asked to take photographs and videos of their experimental studies, and to prepare mini lecture-demonstrations to communicate their projects to the whole classroom. In many cases, one more step was added, namely, the creation of a science-communication scenario, combining the experiments of all separate groups. This final step often sparked the imagination of the students. It resulted in the collaborative preparation of smart theatrical sketches to be presented to audiences wider than the classroom.

4.1 The Student Inquiries: Exploring Different Experimental Traditions

The "Research Creative Activities" course was very successful during the first years of its implementation. During this period, the instructions given by the ministry of education for the organisation and evaluation of this course created a demanding educational environment for both students and teachers. Moreover, 2012–2013 was the last year when groups up to fifteen students were co-ordinated by two classroom teachers. Subsequently, the rule changed so that only one teacher was in a classroom of up to twenty students. Compared to the years before, less teachers were available to stimulate the students' interest and to supervise experimental inquiries.

During 2012–2013, when Panagiotis Balfousias, physics teacher, and I co-ordinated a "Research Creative Activities" course of the first grade, a group of students decided to work on Galileo's experiments concerning the

"WHAT IS HAPPENING IN THE LAB?" 191

fall of objects. Guided by Galileo's texts, and an educational film on Galileo's experiments,[31] they tried to test the fall of different objects in water and air, and the fall of identical objects in different media.[32] They used balls of different materials and weights, and tubes filled with different liquids, such as water, oil and liquid soap (Fig. 10.2, left top and bottom). The students understood very soon that they themselves were part of the experiments because one of the major difficulties was that of releasing the balls at the same time. Furthermore, the short length of the laboratory volumetric cylinders, which determined the distance of the fall in these experiments, caused an additional problem. Although it was easy to observe the balls falling in mediums with augmented resistance, such as the liquid soap or the oil, it was very difficult to observe the balls falling in air or even in water. The students proved very inventive as they tried to examine the simultaneous landing of different balls being released at the same time through the air. They created two electric circuits, each having a small LED lamp which was turned on when the circuit was closed. The circuit had an improvised switch comprising two aluminium sheets placed horizontally, at a short distance from each other, so that they do not touch. When the falling balls landed onto these horizontal aluminium sheets, the sheets touched, activating the switches and lighting the lamps (Fig. 10.2, right). All the experimental trials were videotaped. The analysis of the video-frames offered the possibility of observing the differences in the speeds of the balls falling in the water. It also enabled the observation of the exact moment of the lighting of the lamps, when the balls fell in the air. The contrivance of the improvised switch arrangement was preceded by the augmentation of the distance of the fall. The students let the balls fall to the ground instead of on to the laboratory-bench, increasing the falling distance by about one metre. This alteration failed to satisfy them. There was no significant improvement, and the experiment had difficulties in being videotaped because of the limited free space and unsuitable lighting of the laboratory. These inconveniences sparked-up the idea of the improvised switch, described

---

31 *Six Experiments That Changed the World*, (television documentary mini-series), directed by Bethan Corney and Michael Duxbury, RDF television production for Channel 4, 1999: "Galileo". The film discusses many of Galileo's researches. Among them is the inclined plane experiment, which was relevant here, presented by Thomas B. Settle with the help of a reconstruction.

32 The students were given short excerpts translated into Greek. They carefully worked on Galileo's idea that "the difference of speed in moveables of different heaviness is found to be much greater in more resistant mediums", which was an argument for his thesis that "if one were to remove entirely the resistance of the medium, all materials would descend with equal speed"; G. Galilei, *Op. cit.* (n. 15), p. 75.

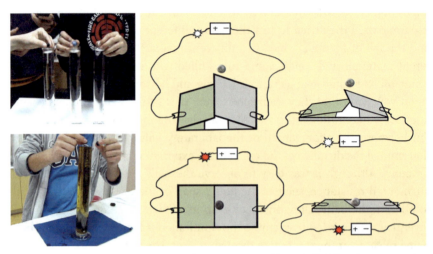

FIGURE 10.2   Experiments on falling objects. *Left top*: Different balls fall in the water; *Left bottom*: Identical balls fall in oil and air; *Right*: Scheme of an improvised switch to check the fall in the air

above, and the students' decision to revert back to the same distance of the fall in all their experiments. The fact that the students recognised as significant problems the synchronisation of the experimenters and the measurement of the time, and searched for ways to confront them, was exciting because these problems constituted significant problems too, in the respective historical experiments.[33]

In the same classroom, another group of students decided to work on von Guericke's and Boyle's experiments with the vacuum pump. Guided by short excerpts of historical documents, they used the vacuum pump to perform experiments, including: the Magdeburg hemispheres, the coin and the feather;[34] the boiling of the water at room temperature;[35] the inflation of a sealed bladder;[36] the gradual disappearance of the sound;[37] and the extinguishing of a flame.[38] Students worked on the presentation of these experiments,

---

33   See: Thomas B. Settle, "An Experiment in the History of Science", *Science*, new ser. 133, 3445 (1961), pp. 19–23. See also: Alexandre Koyré, "An Experiment in Measurement", *Proceedings of the American Philosophical Society* 97, 2 (1953), pp. 222–237.

34   The coin and the feather is a later version of Boyle's experiment on the fall of light objects inside the air-pump receiver. For Boyle's experiment see: Conant, "Robert Boyle's Experiments …", *cit.* (n. 14), pp. 46–47.

35   L. Magalotti, *Op. cit.* (n. 13), p. 57.

36   Boyle, *Op. cit.* (n. 14), v. 2, pp. 412–413.

37   Conant, "Robert Boyle's Experiments …", *cit.* (n. 14), pp. 34–35.

38   Boyle, *Op. cit.* (n. 14) v. 2, p. 517.

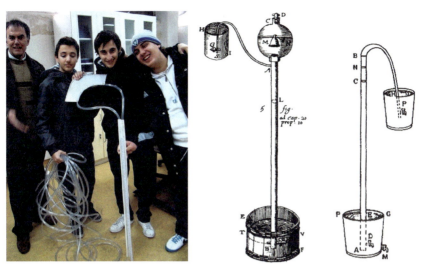

FIGURE 10.3   Water, air and vacuum. *Left*: Tools for replicating Gasparo Berti's experiment; *Right*: Two sketches of Berti's experiment
AFTER EMMANUEL MAIGNAN, *CURSUS PHILOSOPHICUS*, TOULOUSE, APUD RAYMUNDUM BOSC, 1653, AND RAFFAELLO MAGGIOTTI, "LETTER TO MERSENNE", DATED 1648

most of which are really spectacular. They learned to interact with their audience, asking them to become reliable witnesses. They also learned to use discussion and surprise as a means to grab the audience's attention.

During 2012–2013, a third cohort of students from the same classroom decided to work on Berti's experimental idea concerning the possibility of creating a vacuum.[39] The students were asked to begin their project by proving that water does not leak from water-filled tubes of lengths of two metres, and then five metres when these tubes have their upper ends sealed off and are positioned vertically into containers full of water. Experiments with both two-metre and five-metre long tubes proved successful. Next, students were asked to find out what would happen if the same experiment was conducted with an eleven-metre tube. For all their experiments, the students were provided with a flexible transparent vinyl tubing (Fig. 10.3, left) instead of rigid-walled hollow glass or metal tubes, which were historically used in the final version of Berti's experiment (Fig. 10.3, first sketch). Very soon, students realised that it was impossible to seal the eleven-metre tube effectively. Panagiotis Balfousias, who mainly supervised this group, had the idea to recreate the set-up shown in the second sketch of Fig. 10.3 with one alteration. A twenty-two-metre flexible

---

39   Middleton, *The History of the Barometer*, *cit.* (n. 11), pp. 10–18; Middleton, "The Place of Torricelli …", *cit.* (n. 11), pp. 11–17; Shea, *Op. cit.* (n. 11), pp. 24–31.

vinyl tubing had to be used instead of an eleven-metre one, in order to keep the two water containers at the same level. In this way the two open ends of the tubing would stay in the playground and there would be no need to seal the upper part of the tube in order to transfer it to a higher level. Thus, the students filled the tubing with water from the tap in the schoolyard and submerged its two open ends in the same single container full of water. Without removing the tubing ends from the container, the middle of the tubing was draped over a hanger and, with the help of a rope, it was lifted and attached to the exterior wall of a second-floor window of the school. It was very impressive to observe the water bubbling in the upper part of the tubing, and the upper part of the tubing being cooled. This effect was associated with the boiling of the water at room temperature inside the air-pump glass jar, with which the students were already familiar. Then, the students were asked to check the Torricellian theory by investigating if their set-up could be used as a barometer.[40] For the first two days all went well; as the weather got worse, the level of the water inside the tubing dropped a little. Unfortunately, on the following days, the weather improved considerably, but the level of the water continued to drop ... The reason for this effect remained unknown. Some of us supposed that the walls of the plastic tube permitted leakages, others that there was air dissolved in the water of the tube, and a third group that the culprit was the presence of water vapour in the upper part of the tube. These premises were not further checked. Both our success in managing to make the water boil in the upper part of the tube together with our failure to construct a reliable barometer provided the means to comment on the philosophy of the experiment. We explicitly explained to the students the importance of observing and systematically registering all the circumstantial details of an experimental process. The students used this information creatively and recorded through photographs and videos all the stages of their project. They showed great diligence in registering the level of the water inside the barometer-tube once or twice a day. They also understood the significance of unexpected results because they witnessed how specific circumstantial details led to the alteration of the initial set-up idea. Finally, we challenged them to consider the scientific value of inconclusive or failed experimental attempts. We analysed the case of Berti's experiment, which, although it succeeded in causing the bubbling of the water in the upper part of the tube, it remained inconclusive concerning the production

---

40   See: Evangelista Torricelli, Letter to Michelangelo Ricci dated 11 June 1644, in: Gino Loria and Giuseppe Vassura (eds.), *Opere di Evangelista Torricelli*, Stab. Lito-Tip. G. Montanari, Faenza, 1919–1944, 5 vols.: v. 3 (1919), pp. 186–189. See also: Middleton, *The History of the Barometer*, cit. (n. 11), pp. 19–32; Shea, *Op. cit.* (n. 11), pp. 32–34.

"WHAT IS HAPPENING IN THE LAB?" 195

of a vacuum. We also highlighted the case of Torricelli, whose idea for the creation of a barometer proved correct, although his experimental trials for the construction of a barometer did not succeed.[41]

The fourth group of students from the same classroom cohort was involved with static electricity experiments. They were aware of the basic concepts of the theory of electricity from previous school grades. Through the introductory lectures, we drew their attention to the fact that during the eighteenth century, electricity experiments became very popular. Also, that, together with great scientists, a plethora of amateurs and experimenters, who made their living with electricity, contributed to scientific developments by changing and improving the experimental arrangements.[42] Attracted by the Wimshurst machine,[43] the students searched for ways to enlarge the spark produced.[44] They found that the length of the spark could be increased if the Wimshurst terminals were connected to large surface metallic conductors or Leyden jars (Fig. 10.4, left).[45] They examined whether the strength of the spark increased when they filled the Leyden jars with fresh or salted water. They then decided to create a model of electric bells inspired by the "Franklin's bells" experiment.[46] They made an arrangement consisting of a pendulum which swung between two hemispherical conductors adjusted on insulated bases and connected to the Wimshurst terminals. They spent much time improving the whole set-up by choosing the best pendulum bob. Both its weight and the sound produced, when it hit the hemispherical conductors, were examined. As well as this,

---

41   See: *Ibid.*, pp. 34–36.

42   See: Roller and Roller, *Op. cit.* (n. 17), pp. 591–596. See also: Bernadette Bensaude-Vincent and Christine Blondel, "A Science Full of Shocks, Sparks and Smells", in Bensaude-Vincent and Blondel (eds.), *Op. cit.* (n. 29), pp. 1 and 6; Paola Bertucci, "Domestic Spectacles: Electrical Instruments between Business and Conversation", in *Ibid.*, pp. 75–87; Oliver Hochadel, "The Sale of Shocks and Sparks: Itinerant Electricians in the German Enlightenment", in *Ibid.*, pp. 89–101.

43   Our Wimshurst machine, an instrument whose design dates back to the 1880s, was introduced to the students as a representative of the electrostatic generators and as an instrument of the Baconian tradition, which could create artificial lightning. The frictional electrical machines of the 1750s, which were mostly used in the electricity shows, were presented to the students as ancestors of the Wimshurst machine.

44   For eighteenth-century experimenters to strengthen the spark produced by the electric devices was a common problem; see: Roller and Roller, *Op. cit.* (n. 17), p. 595. Remarkably, the students were spontaneously involved with the same problem.

45   The Leyden jars they used were similar to those that were part of the Wimshurst machine. Aluminium foil was used to cover the inside and outside of a cylindrical plastic container. The container was covered with a plastic cap, to which a metallic wire was properly adjusted. The Leyden jars were constructed by the students of this group.

46   See: Franklin, *Op. cit.* (n. 17).

FIGURE 10.4   Static electricity. *Left*: Large surface conductors and Leyden jars to strengthen the Wimshurst machine spark; *Right*: Water-filled Leyden jar

the students tried to find how the whole effect was influenced, when the above-mentioned Leyden jars were added to the entire set-up, and connected to the hemispherical conductors. This experimental inquiry, together with other experiments, such as the bending of a stream of water by an electrified rod, the creation and examination of water-filled Leyden jars (Fig. 10.4, right),[47] the electric kiss,[48] and the electric chain,[49] comprised their lecture-demonstration delivered to the whole classroom. During its preparation, the students were occupied with improving the interaction with their audience, and the organisation of their lecture bench. They were interested in making their experiments clearly observed and most attractive.

During the next year, some of the students of the project on static electricity, together with new partners, decided to work on electromagnetism, within the framework of the "Research Creative Activities" course. Through the introductory lectures, I presented basic instruments and experiments on electromagnetism. Among these were Oersted's experiment, Faraday's researches on electromagnetic induction,[50] Ruhmkorff's induction coil, and Crookes' and Thomson's cathode ray tubes. The laboratory Braun (or cathode ray) tube was presented as an educational tool for the measurement of the e/m ratio of the electron, while the oscilloscopes were introduced as a means to illustrate

---

47   The students checked the operation of a laboratory Leyden jar similar to the one described by Pieter van Musschenbroek; see Roller and Roller, *Op. cit.* (n. 17), pp. 594–595.
48   See: *Ibid.*, pp. 591–593.
49   See: Nollet, *Op. cit.* (n. 17), p. 18.
50   Together with experiments on electromagnetic induction, I showed the film: *Six Experiments that Changed the World*, cit. (n. 31): "Faraday".

FIGURE 10.5   Electromagnetism. *Top left and right*: Writing with electricity; *Bottom left*: Adding a magnetic needle to illustrate the voltage change; *Bottom right*: Lissajous curves with a Braun tube

electrical signals. The students started their experimental studies by repeating the basic experiments on mutual and self-induction and by checking the use of the oscilloscopes and the Braun's tube. They were helped to recreate the arrangements of the experiments they wanted to study. They were allowed to use all the equipment presented in the introductory lectures, except for some very fragile vacuum tubes, such as the Crookes' and Thomson's examples.

After this preliminary stage, the students managed to successfully combine their knowledge with their imagination. They contrived new experimental set-ups. The first student-initiated experiment was a circuit comprising a low voltage battery, a set of coils connected to each other and an improvised manual switch. The switch was a large metallic plate in which an insulated wire connected to the circuit was touched or removed. The electromagnetic induction voltage produced by the quick interruption of the circuit caused a small spark between the wire and the plate. The spark was used to "write with electricity", by moving the position of the spark on the metal plate of the manual switch, and producing oxidation marks on it (Fig. 10.5, top left and right). Thus, an excellent demonstration experiment was created. In the next iteration of this experiment, the same students placed a magnetic needle over the coils, which could rotate freely in the vertical plane defined by the axes of the two coils. Motion of the pivoted magnetic needle was used to show the voltage change (Fig. 10.5, bottom left). This idea of using a magnetic needle to illustrate the voltage change constitutes the initial point for the transformation of the whole construction into a measurement instrument. It was very remarkable in this case that the students managed to combine two different things: a demonstration experiment and an idea for a measurement instrument. In a third experimental set-up created by this same group, a Braun tube was used. However, instead of using it to measure the e/m ratio, they used it to produce Lissajous's curves (Fig. 10.5, bottom right). In this case, an instrument introduced as a measurement device was used to create a demonstration experiment. The fourth experimental set-up created by this group consisted of the successful repair of an electromagnetic generator, which was part of the 1970s instrument collection of the school laboratory and for years was out of order.

The most impressive characteristic of the projects on static electricity and electromagnetism, described above, was the inventive character of the students' research. The students were really attracted by their experimental inquiries and managed to offer their own answers to their experimental questions. They succeeded in improving and completing the experimental set-ups. They proved able to imagine new instruments and experiments. It was also surprising to them that many of their experimental questions, such as the strengthening of the spark or the creation of an instrument to illustrate the voltage change, were also historical scientific questions. There were indeed very talented students in the electricity and electromagnetism groups. And it was a great satisfaction to their teachers when these students confessed that the laboratory projects helped their creativity, because they stimulated their interests and they were allowed to work in group settings. That was the reason that they chose to follow the laboratory courses for two successive years. Some of them were also members of the extracurricular laboratory course, which allowed them to work further on their ideas and develop their experimental set-ups.

4.2   *Lecture Demonstrations and Science Communication Scenarios*

Lecture demonstrations and science-theatre were extensively used as educational tools in our projects. Lecture demonstrations were used to introduce our laboratory courses, while at the same time constituting an initial step in the students' projects. In a further development, my lecture demonstrations linked our laboratory with history of science museums.

During the academic years 2013–2014 and 2014–2015, I was invited to prepare two lecture demonstrations to help the Athens University History Museum make its nineteenth-century scientific instrument collection attractive to

"WHAT IS HAPPENING IN THE LAB?" 199

young audiences.[51] The instruments of the collection should not be touched. Thus, instruments from our school were brought to the museum for the presentation of live experiments, throwing light on the use of the historical objects. The scenarios of the lecture demonstrations, which combined narration, live experiments and audio-visual presentation, highlighted the nineteenth century, the period when the older instruments of the Athens University History Museum were acquired. To be sure that all the members of the audience could observe every single spark, colour, and so on, produced on the lecture bench, I decided to include, in the audio-visual material, videos of the experiments and project them in parallel with their live performance. I had the idea to engage the students participating in the extracurricular laboratory project in the preparation of these videos. Also, I involved my students in the performance of the same experiments in front of the museum visitors (Figure 10.6 bottom). Although the students did not participate in the creation of the lecture-demonstration scenarios, they still enjoyed a rich science communication experience.[52] That personal growth became very clear when they were asked to design their own scenarios, later in the school year.

The contemporary Greek financial crisis inspired the plot of the scenario that students created to present their 2014–2015 extracurricular-laboratory course results. The creative premise of the scenario gave the primary role to a man who had lost everything, and searched for ways to commit suicide. In the plot, suddenly a shop sign caught his attention: "Laboratory of dangerous experiments – Staff required". Being hired as an assistant there, he expressed extreme willingness to participate in the riskiest experiments. Such experiments formed a part of our show: Leyden jar sparks, hydrogen balloons, Volta pistols, and so on. But, even as the students performed these experiments for the audience, the plot presented the new assistant as being barred from performing them, due to his lack of experience. Finally, one day when everybody had left for lunch, he stabbed himself. When his colleagues returned, they

---

51   On the Museum, see: http://en.historymuseum.uoa.gr/ (accessed 7 Nov. 2020). Both lecture demonstrations were incorporated in the science communication programme of the museum titled "Ask the Scientist", which was coordinated by Fay Tsitou. The programme was addressed to high school students.

52   The students had the opportunity to witness different phases of the lecture-demonstration process: how the lecture-demonstration scenario was linked to the museum exhibits; how our experiments highlighted the use of the museum instruments; how different science communication tools, such as live performances of experiments, narration, audio-visual presentation and discussion were combined to catch the audience attention.

FIGURE 10.6   Science theatre and lecture demonstrations. *Top*: Science theatre, 2015; *Bottom*: Lecture demonstrations at the Athens University History Museum in 2014 (*left*) and 2015 (*right*)

found him on the ground, covered in 'blood' (Fig. 10.6, top). After some seconds of suspense, with the assistant still on the ground, one of the actors explained to the audience our trick: "The blood-red coloured liquid on the assistant's wrist was iron III thiocyanate, the result of a chemical reaction between two aqueous solutions, the light orange solution of iron III chloride and the colourless one of ammonium thiocyanate". The actor performed the same chemical reaction in a test tube. The knife, the assistant used, had previously been put in a receiver with ammonium thiocyanate, while he himself, without being

"WHAT IS HAPPENING IN THE LAB?" 201

perceived by the audience, had placed his wrist on a cotton-wool ball completely soaked in iron III chloride.[53] This science show, which lasted about fifteen minutes, and was presented at local level to an audience of students and teachers of different schools, imitated the style of eighteenth-century public shows because it succeeded in communicating scientific information, in satisfying the senses and in attracting the attention of different kinds of attendees.

4.3   *The Escape-Room Project*

The basic idea of the escape-room game, mentioned in the introduction, was to make the participants discover historical experiments with the help of an interactive game. In a preparatory meeting between the two collaborating school-teams held in our school, our laboratory was successively transformed into a museum of experiments and an escape room. During the first part of the meeting, multiple series of experiments on seventeenth- and nineteenth-century researches on air, vacuum, light, colours and spectroscopy were contextually presented by the 1st Lyceum of Vrilissia school-team. Students performed and explained the experiments, while I played the role of the narrator commenting on their history and nature. In the second part of the meeting, the collaborating school-team introduced us to their idea for an escape-room on Galileo. They, firstly transformed the laboratory into a theatrical setting with the help of objects, such as books, paintings, pages of historical texts and pictures, decorated boxes, and experimental set-ups. Then, using verses, word-games, and theatrical sketches, they guided our students to find objects and instructions needed to conduct Galileo's pendulum experiment.[54] The successful completion of this experiment and the finding of the isochronism of the pendulum oscillations permitted the exit from the escape-room.

The result of this meeting was the decision to include in the final escape-room game all of the series of experiments presented, forming the six different rooms mentioned in the introduction. During the next two months, we worked hard to transform historical information into clue-solving processes. This phase stimulated students' interest, which had a great contribution in the weaving of the escape-room scenario. We used two different approaches. The first was to use clue-solving processes to lead the participants to accomplish the experiments themselves. The second was to ask for the active participation

---

53   This science show included experiments that could be dangerous in the hands of inexperienced experimenters. The students were trained to follow all safety precautions. The iron III thiocyanate produced on the assistant's wrist was carefully removed immediately after the few seconds that corresponded to the duration of this scene.

54   See: Galilei, *Op. cit.*, (n. 15), pp. 94–99.

FIGURE 10.7   The escape-room project at the Athens Science Festival 2016. *Above*: Two students interpret John Tyndall and Lord Rayleigh lecturing at the Royal Institution of Great Britain; *Below*: Visitors explore the navigation process

of the visitors in science-theatre sketches imitating historical science communication events. The history of the prism inspired Newton's room scenario,[55] created by the collaborating school. The visitors had to discover a coin on the threshold of the natural magic shop and use it to buy a prism, as Newton did. Then, they had to bring the prism to Newton's room to conduct his experiments on light and colours. The seventeenth-century experimental culture of witnessing sparked up our Royal Society room scenario. An invitation letter dated 5th April 1660 and signed by Robert Boyle and Robert Hooke led the visitors to this room. The letter explained that "crucial experiments, which were to be conducted with the newly invented instrument – the air pump – needed to be observed and reliably witnessed". Two students from our school, who represented the above-mentioned scientists, performed the experiments and asked the visitors to register all circumstantial details and sign off the proceedings, as

---

55   See n. 25.

verification (Fig. 10.1, above). The completion of the proceedings was the key to exit this room. Analogously, the nineteenth-century lectures at the Royal Institution, which were at the same time spectacular, scientifically accurate and referring to the latest contemporary discoveries, inspired the theatrical-game taking place in this room (Fig. 10.7, above), which was described in the introduction. The presence of Sherlock Holmes was a way to highlight the great impact of science in society during this period.

In the Athens Science Festival 2016 escape-room event, students from our school performed or supervised experiments in all rooms. The students from the collaborating school had the leading role in the guidance of the participants from room to room. A clever general navigation plan contrived by the collaborating schoolteam allowed more than fifty participants to be in the escape room at one time (Fig. 10.7, below). Keywords linked to more than one room – such as the words spectrum, mercury, vacuum and prisms – had to be combined to solve the whole escape-room mystery and allow the exit from it.

4.4   *The Deutsches Museum Exploration*

A science communication event was also used as a means to summarise a class visit to the Deutsches Museum in Munich, in March 2017. Before the visit, we spent several months working on historical experiments and science-and-society issues. The students saw different parts of the museum, through photographs and videos, and conducted experiments inspired by them.[56] They were also challenged to reflect on topics such as: science and power; science, technology and society; and science and war, based on works of nineteenth- and early twentieth-century thinkers.[57]

As this museum constitutes an exciting teaching environment, we spent two whole days there. We visited the mines, astronomy, physics, telecommunications, historical factories, and the history of the museum sections. The students were impressed by the flying machines (Fig. 10.8) and the electricity show, enjoyed the live construction of the Crookes' radiometer, were surprised

---

56   We again explored the seventeenth-century research on the vacuum, eighteenth-century electricity, and nineteenth-century electromagnetism. The students created a Faraday's cage apparatus to approach the electricity show of the museum. They also worked on chemistry and astronomy experiments.

57   In order to introduce the above mentioned science and society issues, I used excerpts from the films: *The Invisible Man* (1933), directed by James Whale and produced by Universal Pictures; *The Time Machine* (1960), directed by George Pal and produced by George Pal Productions, Galaxy Films Inc.; *Modern Times* (1936), directed by Charles Chaplin and produced by Charles Chaplin Productions; *Einstein and Eddington* (2008), directed by Philip Martin and produced by Company Television Productions, HBO Films, Pioneer Pictures; and *The Imitation Game* (2014), directed by Morten Tyldum and produced by Black Bear Pictures, Film Nation Entertainment, Bristol Automotive, Orange Corp, The Weinstein Company, Hishow Entertainment, West London Films.

FIGURE 10.8   A view of the Flying Machine section at the Deutsches Museum, Munich (2017)

by the piano that played without a pianist, and stopped in front of Otto Hahn's table.

Back at school, a group of five students of the extracurricular laboratory project undertook the task to work on a mini science-theatre performance in order to present the whole visit experience to the local educational community. The students and I began to create the script by combining pictures from the museum exhibits with live experiments. During this process, our attention became riveted by the image of a notebook cover, found among our archive of hundreds of photographs taken on the trip (Fig. 10.9). Its fine sketches were presented under the title *Les Merveilles de la Science*. All the machines depicted there could be found in the Deutsches Museum rooms. But the notebook owner was not a museum visitor. Our photo was taken of a display that we saw at the Dachau concentration camp memorial, during our short-class visit. The notebook label "Cahier de Ma Vie en Prison", sparked our imagination. Who was the writer? How he lived? We imagined him writing during the nights, and dreaming of his escape with the help of the machines. We imagined him guiding us to the museum with his notebook drawings. We made him the central character in our science communication scenario. But the persistent question concerning the real history of the person,[58] compared to the dreamy thoughts

---

58   Some months after the end of our project, I managed to trace the notebook writer, whose name was Josef Helsen. According to the archives, he was born in Belgium in 1913. He was

"WHAT IS HAPPENING IN THE LAB?" 205

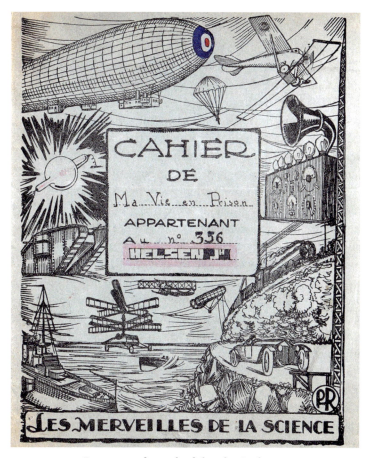

FIGURE 10.9   Front cover of a notebook found in Dachau
COURTESY OF YAD VASHEM ARCHIVES, JERUSALEM

inspired by the notebook cover, was the best illustration we could find for the contradiction between the nineteenth-century science promises for happiness and the nightmarish reality of twentieth-century wars.

transported to Dachau from Natzweiler, in September 1944, and was imprisoned there until January 1945, when he was transported to Mittelbau-Dora concentration camp. He was a Night and Fog Decree prisoner. There is no information about his death. In the notebook, now kept in Yad Vashem Archives in Jerusalem, there are a few pages written in French. Helsen writes to his wife, giving details about his cell companions and his life in the camp. Paragraph after paragraph, the writing becomes sadder and stops suddenly on the fourth page. The calligraphy indicates a high educational level, and a gentle and tender character. The document we saw in the Dachau exhibition was a scan of the notebook cover. A clearer photo, sent to us by the Yad Vashem Archives, shows that the figures of the cover are printed and not painted, as we assumed in the beginning.

5   Conclusions

Through all the examples presented above, I have tried to show that we managed to give the school laboratory its own life, as we transformed it from an auxiliary place for the teaching of science into a place where students actively discover. The same objects, in its shelves, which students used to understand as simple practical tools by means of which experiments are conducted became time machines, inquiry vehicles and magic wands to fascinate audiences. The students' response to our initiative was clearly positive. From 2012 to 2020, there were no more than ten occasions when a session of our evening laboratory course was not held due to the students' absence. The students' interest in this optional course was motivated through our curricular laboratory activities. Although it seemed reasonable that only science-orientation-group students would participate in our extracurricular laboratory programme, the science-and-society and science-communication focus given to our projects functioned as a pole of attraction for students with other interests too. A significant proportion of the students, who comprised the science-theatre team of the 2019–2020 project, were pursuing classics studies. Similarly, one of the five students, who participated in the 2014–2015 science communication activities, followed literary studies.

In many of our laboratory projects, students and teachers worked as equal members of groups engaged in exploring instrument and experiment issues. Although the teachers had the dominating role in introducing the experimental projects, in many cases, the students used the information offered to create their paths of inquiry, by posing and solving unpredictable experimental questions. The initiative role of the students was apparent in the cases of electricity and falling-bodies experimental studies. However, even when the student-groups needed more guidance, it was exciting how positive and diligent they were. In contrast to the typical science courses, where they are asked to conduct part of an experiment or are just taking some measurements, the students learned to develop complete experimental processes. They became able to handle, alter or create their equipment. They learned to conduct numerous experimental trials and register the experimental results in detail. They learned to consider inconclusive or failed experimental attempts as integral parts of experimental research.

Our laboratory projects emphasised science communication issues. The students learned not only to do experiments or create experimental set-ups, but also to use and invent different ways to present their experiments and interact with their audience. The students proved surprisingly inventive in science communication issues. In many cases, they had the primary role in the weaving of scenarios. However, even in the cases that the teachers drew the main axes of the science communication events, the students intervened with their ideas, which could nicely predict how their peers' interest would be aroused. The students were keen on playing the different roles occurring in the scenarios. It was exciting how convincing they were in the role of seventeenth-century presenters of matters of fact, and how successfully they persuaded the audience participants to play the roles of reliable witnesses. Students were equally successful when they acted as eighteenth-century public demonstrators or nineteenth-century lecturers.

Although our students did not have the opportunity to directly handle historical scientific instruments, these artefacts were always present in our projects and understanding. My introductory lecture demonstrations focused on the links between the equipment of our laboratory and historical instruments in museums and collections. The historical scientific instruments became standard references in the students' studies. The students tried to imitate historical experimental set-ups, and experiments or science communication events. In many cases, our projects were combined with visits to history of science museums. We realised that our students were well-prepared visitors to these museums, as they were familiar with Leyden jars and electrostatic machines, with vacuum pumps, Ruhmkorff coils, X-ray set-ups, and many other apparatuses. In Greece, there is a plethora of archaeological museums, but few of historical scientific instrument collections. Through our laboratory projects, the students had the opportunity to interact with the European scientific instrument culture and serve as ambassadors to other students. All the visitors to our escape room event in the 2016 Athens Science Festival learned about the existence of the Royal Society of London and the Royal Institution of Great Britain as historical places of science. The presence of our students, as performers of experiments in the case of the Athens University History Museum lecture demonstrations, served as an incentive for other students to explore the historical scientific instrument collection of the museum. Thus, the historical scientific instruments served as a source for our laboratory projects, and our laboratory projects served as bridges between historical scientific instrument collections and school-students.

Through our laboratory projects, philosophy-of-scientific-instrument issues were brought to light. Our students diversely engaged with these ideas: that there are different kinds of instruments and experiments; that scientific instruments are created for audiences; that historical scientific instruments serve as ways to objectify knowledge; that the making and use of scientific instruments can take unpredictable paths; and that science, society and culture are variously intertwined.

The element of surprise was frequently present in our projects. Sometimes, it emerged from the students' initiatives (as with the electricity experiments or the 2015 scenario) or the unexpected success of science communication events (as with the 2016 escape-room project). In other cases, it made its appearance through unpredictable experimental incidents (as with Berti's experiment reconstruction), or strange encounters with history (as with the Deutsches Museum visit). All these small revelations indicate that our courses were not only the simple repetition of a good recipe, but a step towards the visualisation of a new laboratory culture: a culture able to combine the past with the present, and the different interests, skills and talents of the students with the multifaceted approach of scientific objects via history, philosophy and society perspectives.

Acknowledgements

I would like to express my very great appreciation to Elizabeth Cavicchi, Peter Heering and Giorgio Strano for their valuable comments during the reviewing of this paper. I must also thank the organisers and participants of the question-and-answers SIC 2020 Teaching Session for the constructive discussion concerning the introduction of historical scientific instruments in education. I am also extremely grateful to the Athens University History Museum for the opportunity it gave me to participate in their programmes as science communicator. I sincerely thank the Dachau Museum and the Yad Vashem Archives of Jerusalem for the information provided. I would also like to thank the Yad Vashem Archives for the permission to publish the photograph of the notebook found on display in Dachau. I deeply thank the teachers of the 4th Lyceum of Palaio Faliro, Vasiliki Milioni and Polyxeni Simou, and the escape-room student team of this school for the exciting educational experience we shared. I offer my special thanks to the teachers from my school who I collaborated with, as well as to my students. Their willingness to learn, share and participate always constitutes the core of my work.

CHAPTER 11

Historical Instruments, Education, and Do-It-Yourself in the Cabinet of Curiosity of Brest, France

University Experiences in Mathematics

Frédérique Plantevin and Pietro Milici |
ORCID: 0000-0002-0079-6702; 0000-0003-0478-6306

1 The Cabinet of Curiosity of Brest

The University of Brest, now called Université de Bretagne Occidentale (UBO), is relatively young: founded as a satellite of Rennes' Faculty of Science in 1963, it became a fully autonomous university three years later. Almost another fifty years from then, a former chemistry laboratory was transformed into a place where scientific instruments are being collected to be organised, explored, exhibited, and manipulated: the Cabinet of Curiosity (Fig. 11.1).[1] Naturally, the collection of these devices began long before.

1.1 *IREM Collection*
From its beginning, the Faculty of Science hosted an Institute for Research in the Teaching of Mathematics (IREM), designed to manage the continuing training of mathematics teachers through the use of pedagogical materials, from electronic calculators to geometric devices (Fig. 11.2).

IREM's first introduction of obsolete instruments in mathematics classrooms dates back to the 1990s when slide rules collected among university teachers and engineers were shown to students to encourage them to revisit real/decimal numbers, operations, and logarithmic functions. Enthusiastic feedback made the choice of mathematical instruments grow continuously: the collection expanded by acquisitions and partnerships with private collectors. Such artefacts were introduced in teaching experiences with primary and secondary

---

1 This chapter is based on the presentation: Frédérique Plantevin, "Historical machines, education and DIY in the Cabinet of Curiosity of Brest, France", *XXXIX Symposium of the Scientific Instrument Commission*, London , UK, 14–18 September, 2020. For an overview of the Cabinet, see: www.univ-brest.fr/patrimoine-scientifique/Patrimoine+Scientifique/Cabinet-de-curiosite/Presentation (accessed 20 Oct. 2020).

© FRÉDÉRIQUE PLANTEVIN AND PIETRO MILICI, 2022 | DOI:10.1163/9789004499676_013

FIGURE 11.1   The Cabinet of Curiosity of Brest
PHOTO AND INFOGRAPHICS BY MARC DANAUX, UBO

FIGURE 11.2   Some mathematical instruments of the Institute for Research in the Teaching of Mathematics (IREM) collection: pantographs (bought and made by students), Napier's and Genaille's rods for multiplications, planimeters to measure the area of a figure

FIGURE 11.3   Posters of the exhibitions loaned by the Musée des arts et métiers (MAM, *left*) and the Science in the Seine and Heritage Association (ASSP, *centre*), or produced by IREM (*right*), that contributed to enlarge collections and deepen didactical practices among academics

teachers involved in specific IREM working groups. In parallel, the first author organised exhibitions between 2008 and 2012 (Fig. 11.3),[2] which included practical workshops for pupils. The displayed instruments and the ones used in workshops were bought (new and used planimeters, for instance), explicitly built for workshops (as Napier's and Genaille's rods), or loaned to UBO by institutions and private collectors, such as the Musée des arts et métiers (MAM) and the Association Science en Seine et Patrimoine (ASSP).[3]

1.2   *The Cabinet's Beginnings*

In the 1960s, the Faculty of Science (at that time, the Scientific University College) had to acquire materials to equip its laboratories for research and teaching in various sectors. Since the 1990s, obsolete equipment (together with other *ad hoc* devices built by university engineers for service reasons or demonstrations) had been organised to constitute a starter collection of scientific instruments. But it was only at the end of 2015 that the first author,

---

2   *Venez prendre l'aire à Brest !*, Bibliothèque Universitaire du Bouguen, Brest, 3 February–12 April 2008 (www.univ-brest.fr/irem/menu/Actions/Expositions+/Exposition-2008.cid169223); *La science nautique au XVIIIe siècle* (itinerant exhibition), 2009 (http://assprouen.free.fr/denoville/exposition.php); *Multipliez !*, Bibliothèque Universitaire du Bouguen, Brest, 7 February–23 June 2012 (www.univ-brest.fr/irem/menu/Actions/Expositions+/Exposition-2012.cid169221, all websites accessed 20 Oct. 2020).

3   For instance, see Claude Cargou, Marie-Paule Cargou and Frédérique Plantevin, *Multipliez ! Instruments de calcul de la multiplication : 200 ans de traits de génie, 200 ans d'industrialisation*, IREM, Brest, 2012.

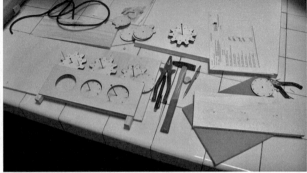

FIGURE 11.4   Examples of works in the Cabinet of Curiosity: navigation tools (*left*) and adding machines (*right*)

supported by colleagues from the collection of scientific instruments, succeeded in founding the Cabinet of Curiosity. This Cabinet combines the scientific instrument collection with the IREM one. In this way, the Faculty of Science got its small informal science museum. The original room design of a laboratory did not change, so that, in the Cabinet, it is possible to touch instruments and see them in action. Furthermore, it has been equipped with do-it-yourself (DIY) tools and materials (cardboard, wood, nails, tapes, strings, etc.) to build new instruments.

From its beginning, the Cabinet hosted various experimental activities on instruments, as IREM working groups and teacher traineeships (Fig. 11.4). In this favourable place, instrument experiences took root, as spread by publications and communications.[4]

As experiences accumulate, ideas about why and how to work with instruments were clarified enough to conceive suitable university courses, namely for students who want to become teachers. Several attempts at creating courses that use these historical instruments in instruction have been made at bachelor's and master's degree levels: each time, instruments built for courses or traineeships have been kept and added to the Cabinet, enlarging and diversifying the mathematical collection.

---

4 Like the national inter-IREM project "Passerelles", where the IREM group "Instruments in History, Instruments in Classrooms" proposed an original work on mechanical calculators: F. Plantevin et Al., "La mécanisation du calcul", in Marc Moyon, Dominique Tournès (eds.), *Passerelles : enseigner les mathématiques par leur histoire au cycle 3*, ARPEME, Paris, 2018, pp. 64–91.

2 Mathematical Instruments in a University Course

In what follows, we focus on a specific course offered for third-year bachelor students of mathematics: the 3-ECTS course *Ouverture vers le master* (Opening towards the master). The innovative course aims to make students – mainly future mathematics teachers – experience activities by linking mathematical thinking and instruments in ways that enrich their mathematics ideas from a material, pedagogical, historical, and epistemological perspective.[5] We next describe the five goals that we developed for the course.

2.1 *Goals of the Course*

The use of instruments in sciences is at the basis of the experimental method. However, the relationship between mathematics and the real world is less direct than for the other sciences. Actually, for many professional mathematicians, the beauty of mathematics is given by its perfect ideality, far from its relation with the imperfect real world: that would make mathematics more related to philosophy than to the sciences. Therefore, our first goal is to experience the materiality of theoretical concepts by linking instruments to mathematical contents, both in the analysis (to understand the behaviour of the real artefact and its causes) and in the synthesis (when modifying, designing, or building a real machine implementing a mathematical idea). Although academics and sometimes authorities recommended experimental mathematical activities with instruments, their actual educational use for manipulative activities remains unusual,[6] often bound to associations like the *Laboratorio delle Macchine Matematiche* (Mathematical-Machines Laboratory) of Modena, Italy.[7]

Some artefacts carry specific historical backgrounds that can highlight a setting, or a mathematical foundation, and express properties that are different from those typical of today's mathematics. Study of these artefacts could improve students' critical thought and add a degree of historical heft to their education, providing a glimpse of the long and rich process of building mathematical ideas. For these reasons, we chose characteristic historical artefacts to reach our second goal: that of situating artefacts in their original context,

---

5 In Brest, during the bachelor in mathematics, students have no course in history and philosophy of science (neither general nor specific on mathematics), in pedagogy or in didactics.

6 Montessori pedagogical tools for kindergarten and primary schools are well known exceptions.

7 See: www.mmlab.unimore.it/site/home.html (accessed 20 Oct. 2020).

while also considering the mathematics of that time. It is an opportunity for students to get a historical perspective.

Understanding mathematical ideas is a process in evolution: adapting their meaning in new situations is essential for the progress of an individual's knowledge. Jean Piaget described this process of "assimilation and accommodation" as the mechanism by which we pass from a stage of lower knowledge to one of deeper knowledge.[8] Following him, we claim that exploring an instrument, as a new situation in mathematics class, is conducive to generating new representations of the related mathematical concepts. Our third goal is to make students experience an epistemological change of a mathematical concept that is mediated by an instrument.

For students potentially interested in becoming teachers, it is essential to reflect on pedagogical activities that could enrich their future pupils. Therefore, after experiencing first-person activities with instruments, the students are asked to think as teachers: they have to reflect on these activities' potentials for the first time in their education. For instance, they have to set the instruments in suitable grade levels to mediate specific mathematical contents and propose a pedagogical approach introducing such topics.

Besides the other previously introduced goals concerning "hard skills", the last one is to develop students' autonomy.[9] Specifically, although we designed the proposed activity to improve many students' "soft skills" (team working, communication, problem solving), the main aim is to develop their skill of autonomy as mathematicians. This competence, required in master studies, is often disregarded during the bachelor programmes. Hence, as a significant educational aim, the course proposes to facilitate students' development of autonomy by encouraging their freedom, responsibility, and creativity. In acting on this objective, the course breaks the usual schema of frontal lessons, where the teacher lectures from the front of the class. Work with instruments allows us to introduce new activities to students, in a setting which can recall an "a-didactic situation".[10] In our case, students' autonomy is not provided by teacher's requests but by the (autonomous) quest for a conscious exploitation of the instrument.

---

8   See: Jean Piaget, *L'équilibration des structures cognitives : problème central du développement*, PUF, Paris, 1975, pp. 12–16.

9   See: J. Piaget, *Le jugement moral chez l'enfant*, PUF, Paris, 1992, pp. 261–330.

10   According to Guy Brousseau, *Theory of Didactical Situations in Mathematics*, Springer, Dordrecht, 2002, pp. 29–31, a "didactic situation" is an intervention proposed by a teacher to make students construct a certain knowledge. For an effective learning, teachers can design a situation in which the content is not made explicit to learners: this is an "a-didactic situation".

2.2 Course Description

The course comprises five sessions of two hours: two sessions of introductory frontal lessons and three of supervised work in the Cabinet of Curiosity. The final examination consists of a small-group presentation of about twenty minutes, followed by questions. Specifically, characterising aspects are:

– the frontal teaching (kept minimal with respect to the total student activities for this course) introduces the setting and provides a bibliography;
– students have to work in small groups;
– students can choose the subject of their study;
– the manipulation of a real object is highly suggested (though not mandatory), and it is also possible to construct new samples;
– a particular emphasis is given to geometrical instruments, due to their richness from a historical and epistemological perspective;
– the final project consists of a presentation that deepens the instrument's history and proposes a possible pedagogical application of it in school.

While the introduction changed slightly over the years (from 2017 to 2020), it always included epistemological, historical, and pedagogical standpoints and terminated with suitable bibliographical references. From a historical and epistemological perspective, the aim is to give – in a few hours – a minimal introduction to the problem of "exactness", that is (as briefly deepened in *Section 3*), the role of ideal instruments in the foundation of geometric constructions. We chose such a topic because it introduces the foundational importance of instruments in mathematics and is a perspective quite far from today's mainstream approach to mathematics.

After two lessons of frontal teaching, the course meets in the physical space of the Cabinet. A session is devoted to its (partially) guided exploration, and some bibliographic references are provided. Besides original historical instruments, students are introduced to replicas or reconstructions from textual sources and DIY tools made from cardboard and wood. Students can ask for explanations regarding machines and bibliography, divide themselves into small groups (two to four people) and choose a historical mathematical instrument that they will explain in their final examination presentation. Student groups may select for study an instrument which is already present in the Cabinet of Curiosity, or opt for one that is only described in the bibliography.

During the course's final two sessions, work continues in small groups with the teacher's supervision. The teacher can help groups organise the work, analyse the instruments, and realise their DIY constructions (Fig. 11.5).

The final project consists of a presentation in front of the class about the chosen instrument, its analysis, and its possible use in school (adapted to the practices of French academic programmes): using a real instrument and some

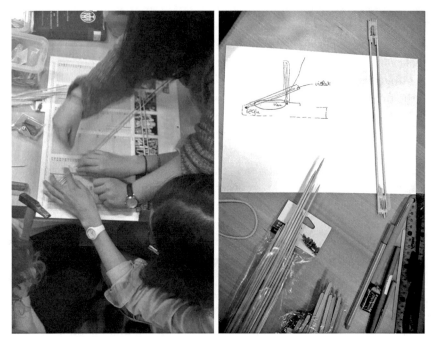

FIGURE 11.5 Different steps of the building processes in the Cabinet of Curiosity; study of books, attempts to understand instruments with a sketch and makeshift equipment

historical information is warmly welcomed. The presentation format is typical for professional conferences: after the twenty-minute slot for the presentation, a questions-and-answers session follows. Both teachers and other students can ask for clarifications, express comments or criticisms.

3   Historical and Epistemological Introduction

Although geometry is merely considered a visualisation tool today, from classical Greece to the seventeenth century geometric constructions were essential to justify the existence of mathematical objects. From this perspective, it was necessary to define the class of instruments acceptable in geometric constructions: this is the problem of "exactness". In this context, instruments played a foundational role that today is no longer taught: these historical mathematical instruments offer a setting where the concept of the curve is treated in a way new to modern students. A curve is not merely a set of points, but makes a continuous trace; it is geometric rather than analytic or numeric.

HISTORICAL INSTRUMENTS, EDUCATION, AND DO-IT-YOURSELF   217

To have an idea of the exactness problem in antiquity, we can consider Euclid's geometric operation as the base of his planar constructions in the first book of the *Elements*.[11] The allowed tools were the formalisation of what today we call straightedge and compass. Still, the hypothetical compass differed from the divider compasses because it could only trace a circle (with a given centre and passing through another given point) without preserving and transporting a given length. The second and third propositions of the *Elements* prove that any divider-compass construction was obtainable by only straightedge and collapsible compass. Euclid's commitment to keep his constructive axioms/ tools as simple as possible (while keeping adequate constructive power) exemplifies what we term an exactness problem in mathematics.

But the limits of Euclid's constructions seemed too strict to solve problems such as the trisection of an angle, the duplication of a cube or the quadrature of a circle. As shown in the eighth lemma of Archimedes, if we assume the use of a marked straightedge (*neusis* constructions), it becomes easy to trisect any angle. But the *neusis* can also be used to construct new curves, like the conchoid of Nicomedes, able to solve the cube's duplication. Therefore, after Euclid, the definition of legitimate construction tools was not so clearly delimited: a quite widely accepted answer came from Pappus of Alexandria (fourth century AD). He proposed three kinds of constructions, of increasing power and decreasing elegance: *planar* (unmarked straightedge and collapsible compass, *i.e.* Euclid's tools), *solid* (besides circles and segments, one can introduce conic sections), and *linear* (general curves can be constructed by tools beyond the previous ones, as the *neusis*). According to the specific problem, a geometer had to use the adequate class of constructions, thus did not have to introduce conics for problems that could be solved (even though with more difficulties) only by lines and circles. About linear constructions, we can note that all the curves known in antiquity were defined by some specific constructions and were only these few: conics, spirals, quadratrices, conchoids, and cissoids.

Pappus's classification constituted the starting point of the ground-breaking Descartes's *Géométrie* (1637), which provided a new canon to define the legitimacy of curves.[12] From one side, Descartes proposed some criteria to define ideal machines to trace curves (synthetic part, construction by continuous motion); from the other, he proposed polynomial equations as a suitable language to study such curves (analytic part, algebraic formulas). The description

---

11   For ancient mathematical concepts discussed in this and the next paragraphs, see Thomas L. Heath, *A History of Greek Mathematics*, Dover, New York, 1981, 2 vols., passim.

12   See: Henk J.M. Bos, *Redefining Geometrical Exactness: Descartes' Transformation of the Early Modern Concept of Construction*, Springer, New York, 2001, pp. 400–401.

of curves by equations introduced the possibility of considering curves thanks to their symbolic definition, without requiring that the curves were constructed or constructible by Euclidean or other subsequent drawing construction tools. That opened the road to an infinity of new curves that can be studied with standardised algebraic methods. In contemporary terminology, we say that Descartes expanded the category of acceptable curves, to include the algebraic ones.

Alongside its enormous successes in many mathematics areas, the Cartesian method was not able to solve two geometrical problems: the classical problem of finding the area under a curve and the modern (seventeenth century) problem of solving the inverse tangent function. Unlike the direct tangent problem (*i.e.* finding the tangent to a given curve), such an inverse problem involves finding the curve whose tangent has to satisfy certain given conditions. The first continuous construction of a curve by inverse tangent conditions was the tractrix, traced by Claude Perrault in Paris (1676) by a pocket watch when dragging its chain's free extremity. The clock's friction on the base plane puts the chain in traction and so tangent to the curve described by the watch. That marked the beginning of a new method for geometrical constructions, called "tractional motion". To overcome the Cartesian canon, Leibniz developed the infinitesimal calculus for the analytical part and used the tractional motion for the synthetic part.[13]

At the beginning of the eighteenth century, some tractional devices overcame the realm of theoretical ideas and became working instruments, mainly thanks to the introduction of a wheel instead of the dragged heavy load (the watch in Perrault's construction). Such a technical improvement allowed the design of quite precise devices for the tractrix and logarithmic curve.[14] However, these devices gradually became forgotten; since the 1750s, the exactness of geometric constructions was no longer emphasised in mainstream research programmes for the arithmetisation of mathematical foundations.[15]

---

13   See: Viktor Blasjo. *Transcendental Curves in the Leibnizian Calculus*, Academic Press, Duxford (UK), 2017, pp. 112–114.

14   See: Davide Crippa and Pietro Milici, "A Relationship between the Tractrix and Logarithmic Curves with Mechanical Applications", *Mathematical Intelligencer* 41 (2019), pp. 29–34.

15   Carl B. Boyer, *A History of Mathematics*, John Wiley & Sons, New York, 1968, pp. 598–619.

4 Students' Production

Although students were free to choose the instruments and subjects of their study in the course, we proposed some possible works as historical introduction. Specifically, the proposed bibliographic references are organised into four subject groups:

- Devices for constructions and transformations (such as Scheiner's pantograph, proportional compass);[16]
- Devices tracing conics sections (parabolas, ellipses, hyperbolas);[17]
- Tractional motion (tractrix, logarithmic curve);[18]
- Integral calculators (planimeters).[19]

Some samples of these devices were present in the Cabinet, while others were only described in the resource bibliography that we provided to students.

4.1 *Two Examples*

To provide an idea of the students' works, we present two examples: conics sections and tractional motion. In both these cases, students worked only with bibliographic references relevant to the device of their interest (Fig. 11.6). Students naturally felt the need to build the device, even without a physical model, and to make it work for real.

A group of two students selected to focus on conics. Starting from Camille Lebossé and Corentin Hémery's twenty-first lesson,[20] the student group studied a method to continuously trace a parabola, together with pointwise constructions and the geometric characteristics that define it.

To understand the parabola construction, students found two main difficulties. They had to understand the definition of the parabola by focus and directrix, whereas French students encounter the parabola only as the graph of square functions. Students confronted a further challenge in considering the role of the rope as a geometrical tool, which keeps its length constant.

---

16 Evelyne Barbin, *Les constructions mathématiques avec des instruments et des gestes*, Ellipses, Paris, 2014, pp. 27–56.

17 Camille Lebossé and Corentin Hémery, *Géométrie*, Nathan, Paris, 1961, pp. 271–373.

18 Dominique Tournès, *Construction tractionnelle des équations différentielles*, Blanchard, Paris, 2009, pp. 51–58.

19 John Sang, "Description of platometer", *Transactions of the Royal Scottish Society of Arts* 4 (1852), pp. 119–129; William Thomson, "An integrating machine having a new kinematic principle", *Proceedings of the Royal Society* 24 (1876), pp. 262–265.

20 See: Lebossé and Hémery, *Op. cit.*, (n. 17), pp. 271–373.

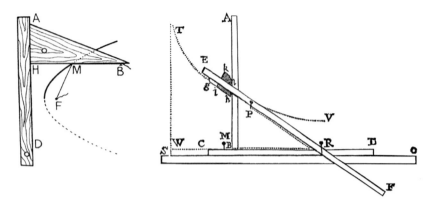

FIGURE 11.6   Two machines represented in the references provided to the students. *Left*: the "parabolograph" to continuously draw the parabola; *Right*: Perks' machine for the quadrature of the hyperbola

Tractional motion was chosen by a group of three students interested in building a device for the exponential function (historically introduced as a logarithmic curve) starting from a slightly modified version of the Perks' instrument for the quadrature of the hyperbola.

For this example, the exponential curve is defined by the geometrical property of keeping constant its subtangent (a definition unknown to students). In doing that, besides understanding the instrument's degrees of freedom, the hardest part is to understand the wheel's role for geometrical constructions, linking its direction to the tangent. As introduced in *Section 3*, this implementation of the inverse tangent problem is related to deep epistemological issues.

In both cases, there emerged the problem of imagining the motion of the instruments by a thought experiment. That imaginative exercise constituted the main push, on the part of the students, to take on the challenge of building real instruments to construct the curves.

4.2   Constructed Devices

For the parabolograph (Fig. 11.7),[21] students considered a rectangle SQRA (working as a set square) sliding along AR. Then they nailed the two extremities of the string of length QR in a point F and in the top point Q of the rectangle. Paying attention to keeping taut the string while making the set square translate, the curve traced by a pencil P along the set square is a parabola of

---

21   See also: www.univ-brest.fr/instruments-scientifiques/menu/Reconstructions+et+Proto types/Parabolographe (accessed 20 Oct. 2020).

HISTORICAL INSTRUMENTS, EDUCATION, AND DO-IT-YOURSELF 221

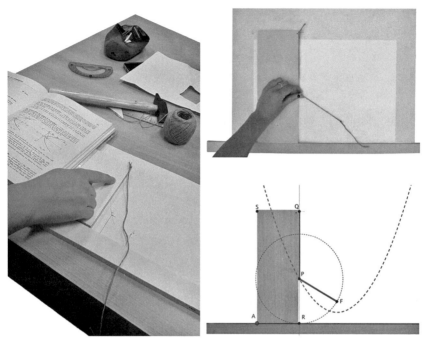

FIGURE 11.7   Realisation of a "parabolograph". *Left*: preliminary studies; *Top right*: the instrument; *Bottom right*: computer simulation

focus F and directrix AR. The parabola is produced because, due to rope length being fixed at QR, P remains at equal distance from F and the line AR.

For the exponential (Fig. 11.8),[22] the curve is traced imposing a constant subtangent. Specifically, a bar (represented horizontally) is fixed on the plane, and an L-rod slides on it (able to move back-and-forth, right and left). As visible in Fig. 11.8 (second detail), a diagonal rod ending with a semicircle and a small sharp wheel (touching perpendicularly the horizontal table surface) is superimposed so that the semicircle touches the L vertical part. Furthermore, a string keeps the diagonal rod close to the nail on the L horizontal component (Fig. 11.8, first detail). Keeping the wheel pressed on the plane, one has to move the diagonal rod toward the left: the apparatus produces a displacement downwards along the vertical part of the L. In this way, as represented in Fig. 11.8 (top right), the wheel traces a (dashed) curve whose (dotted) direction is always given by the diagonal rod. Therefore, such a trace is an exponential curve.

---

22   See also: www.univ-brest.fr/instruments-scientifiques/menu/Reconstructions+et+Proto types/Machine-pour-la-courbe-exponentielle (accessed 20 Oct. 2020).

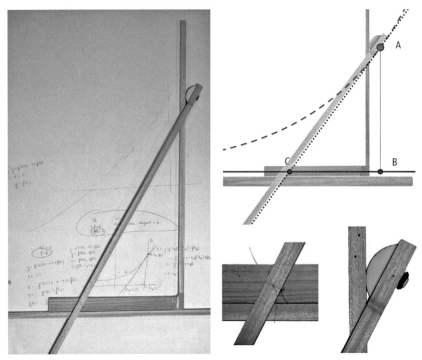

FIGURE 11.8   The exponential machine. *Left*: the instrument; *Top right*: computer simulation evincing the constant subtangent BC; *Bottom right*: two details of the string on the nail, and of the end of the oblique rode with semicircle and wheel

4.3   *Feedbacks*

After the final presentation, we asked the two groups of students for feedback about their feelings and thoughts.

Concerning motivations, students appeared enormously excited by many unusual aspects of the activity. The relation between the course goals presented in *Section 2* and student interests may constitute a way to elucidate their feedback responses and overall excitement. For the first time in their mathematical experience, students had the opportunity to build something tangible with their hands (*materiality*), in freedom, and to take responsibility for it (*autonomy*). Also, the change of perspective played an important role: the possibility of learning something new about well-known subjects provided students with the experience of discovering these mathematical subjects for the first time (*epistemology*). Finally, they were enthusiastic about adopting this kind of laboratory activity into their future classroom activities and teaching (*pedagogy*).

This feedback from the students demonstrated the importance of freedom, practical experiments, and self-made constructions as a motor that propels students' involvement in the activity. Students also appreciated how mathematical devices provided a different perspective on topics well known from a purely modern mathematical standpoint. On the other hand, the instruments' historical background did not constitute an additional interest to motivate students.

In regard to the results that students achieved, both groups said that, after the activity, they felt the traced curves had become more concrete and familiar because they now had an instrument to draw them, analogous to drawing circles with a compass. They meant that the geometrical nature of these curves is strengthened by the existence of real, though imperfect, devices to trace them. Furthermore, this alternative approach enriched their conception of a function graph, from a pointwise plot to a continuously traced curve (epistemological change). To understand this link between instruments and geometrical objects, they had to internalise and "mathematise" some mechanical behaviours, like those of ropes or wheels. The history of the exactness of geometric constructions constitutes a rich and significant set, offering lots of such intriguing examples.

Finally, to understand the device, students had to dive into the mathematical settings at the historical time of the instruments. By doing that, they learned characterisations different from today's approach. In the first example, students experienced the parabola's focus-and-directrix definition, which disappeared from French high-school education (parabolas are presented as graphs of square functions). In the second example, students had to revisit the exponential function as the curve with constant subtangent: by the real experience of tractional motion – theory utterly new to them – they engaged with the non-algebraic nature of the exponential curve and its difference from algebraic curves. In an intertwined way, these results realised our material, historical and epistemological goals.

Students expressed that their perception of mathematical instruments changed (previously, they knew only ruler and compass or electronic calculators). Besides the historical role of instruments in mathematics foundations, students experienced ways that mathematical instruments embody ideas as well as performing tasks. Their future-teachers' toolbox was enriched by these realisations. The simple materials used for the constructed devices reinforce this aspect of relevance to classroom teaching. Everyone can easily build and manipulate these devices (even without access to specific collections). Thus, future teachers can imagine proposing similar activities suitably designed for specific curricular contents to their future pupils, as envisioned by our course's pedagogical goal.

For the construction process, students started from sources not explicitly made for didactical experiences. They had to observe, think, try, and learn from their errors; for example, to understand the exponential machine, they previously used a pencil instead of the wheel (Fig. 11.5). In doing these projects, students took initiatives, constructed connections between material experiments and mathematical knowledge, and engaged with related critical processes of determining relevance and correctness. All these ingredients empowered students to act as independent mathematicians, probably for the first time in their education. This activity fulfilled its aims of fostering students' personal growth in their autonomy as scientists.

5 Conclusions

By analysing these students' mathematical experiences with instruments, we unravelled multiple aspects of our *a priori* motivations in founding these experiences and confronted these motivations with the students' feedback as discussed above. We achieved a list of five main goals (introduced in *Section 2*), guiding our mathematical activities in the Cabinet of Curiosity. These goals provide a framework to evaluate the achieved results and improve the designs of new activities. We hypothesise that such a framework can be adapted to other mathematics courses (for students who do not intend to become teachers, in different grades of education, for teachers), other instruments, and other disciplinary fields. This adaptability also relies on the mutual independence of these goals: meaningful activities may concern only some of them. For instance, the activity with calculating instruments for primary students[23] was designed to pursue autonomy by real open experiments with a strong material component, but did not aim at epistemological changes (classical mechanical calculators are based on today's conceptions of numbers and operations). Moreover, other activities can also naturally introduce interactions between mathematics and other sciences.[24]

---

23   Cf. note 4, and Marc Moyon, Renaud Chorlay, Frédérique Plantevin. "Enseigner les mathématiques par leur histoire au cycle 3", in Fabrice Vandebrouck and Bertrand Lebot (eds.), *Mathématiques au Cycle 3: Actes du Colloque du Plan National de Formation*, IREM, Poitiers, 2018, pp. 87–109.

24   Like calibrating and graduating handmade measure devices or understanding graduations of manufactured ones (Fig. 11.4). We experienced such an activity, between mathematics and physics, when students were graduating quadrants and Jacob's staffs (to measure angular heights) by considering what they are intended to measure and how they run.

In France and many other countries, educational policies are instituted and adopted as attempts to improve students' result in science and, in particular, in mathematics. We believe that our experiments in the Cabinet of Curiosity can contribute towards this effort for they create meaningful activities that enlarge the capacity to reinvest knowledge in concrete situations, reinforce autonomy, and develop soft skills. Furthermore, our emphasis on historical aspects can constitute a strong enrichment to the learners' scientific and mathematical thought, as called for by the ongoing reform of the French high school (history of mathematics is explicitly introduced in curricula).

CHAPTER 12

Educational Experiences in Re-enacting Historical Experimental Procedures

Peter Heering | ORCID: 0000-0002-4002-4168

1 Introduction

The reconstruction of historical instruments has one major basis in the understanding that learners can not only benefit from comprehending science through its history, but also that practical experiences can play an important role in this respect.[1] Obviously, most historical instruments are not suited to be used by novices in the field due to the risk of damage. Consequently, the use of reconstructed instruments can offer appropriate educational opportunities that enable the learner to develop an understanding whilst the historical artefact is not threatened. In this paper, I am going to discuss two case studies: the first one aimed at enabling visitors to a science centre to carry out eighteenth-century electrical salon experiments; the other one addressed to secondary school students, who had the opportunity to build their own version of a historical instrument. Both approaches were evaluated; the results of these evaluations will be sketched.

2 The Electrical Salon: Science Education in a Non-Formal Setting

The first example was based in the Phänomenta, the Flensburg science centre. Phänomenta is one of, if not the oldest, science centre in Germany; it had

---

1 Central aspects of this paper were presented in my paper "Creating and Evaluating Experiences in Re-Enacting Historical Experimental Procedures" at the *XXXIII Symposium of the Scientific Instrument Commission*, Tartu, Estonia, 25–29 August 2014. There are several contributions which emphasise the role of instruments and experiments in science education, see for example: Elizabeth Cavicchi, "Experiences with the Magnetism of Conducting Loops: Historical Instruments, Experimental Replications, and Productive Confusions", *American Journal of Physics* 71, 2 (2003), pp. 156–167; E. Cavicchi, "Learning Science as Explorers: Historical Resonances, Inventive Instruments, Evolving Community", *Interchange* 45, 3–4 (2014), pp. 185–204; Peter Heering, "Getting Shocks: Teaching Secondary School Physics through History", *Science and Education* 9 (2000), pp. 363–373.

© PETER HEERING, 2022 | DOI:10.1163/9789004499676_014

about 3,500 square metres of space, where about 150 experimental set-ups were accessible. Phänomenta had about 70,000 visitors per year, a substantial part being school groups from primary and lower secondary schools. Characteristic to the educational approach of the Phänomenta is the very high degree of de-contextualisation: there are neither written instructions nor explanations in the exhibition. You may find a sign at a set-up, asking a one-line question that serves as an inspiration for the visitors' experimentation. Along with the regular programme, there are always special exhibitions and activities.

From 2012 onwards, we have had the opportunity to set up a historical contextualised room of some 50 square metres for about six weeks. This room was in the upper level of the Nordertor, the historical gate of the city that was built in the late sixteenth century and is not only the town's landmark, but also part of the Phänomenta building. The rooms in the Nordertor are characterised by a plank floor and white brick walls with a slightly irregular surface.

In this room, we re-created electrical experiments from the second half of the eighteenth century.[2] For this purpose, we used a number of instruments that were reconstructed according to historical source information such as an electrophorus, resin cakes for producing Lichtenberg figures, electroscopes, Leyden jars, and entertainment devices such as the electrical boxers and the electrical hailstorm (Fig. 12.1).[3]

Two devices had to be modified compared to the instruments that were used in the eighteenth century: a central instrument in the exhibition was a plate electrical machine, an electrostatic generator in which a glass disc rotates between two cushions covered with cotton fabric (Fig. 12.2, top). The charge from the disc is removed by means of a suction comb and collected on

---

2 For the historical background of these experiments, see in particular John Heilbron, *Electricity in the 17th and 18th Centuries: A Study of Early Modern Physics*, University of California Press, Berkeley, 1979. On itinerant lecturers and their electrical demonstrations in this period see also Oliver Hochadel, *Öffentliche Wissenschaft: Elektrizität in Der Deutschen Aufklärung*, Göttingen, Wallstein, 2003. This electrical salon can be seen as a further development of one part of an earlier exhibition I co-curated at the Museum Mensch und Natur in Oldenburg: *Welt erforschen – Welten konstruieren: Physikalische Experimentierkultur vom 16. bis zum 19. Jahrhundert* (Exploring the World, Constructing Worlds: Experimental Cultures of Physics from the 16th to the 19th Century). On this exhibition, which addressed historical experimental practices, see Peter Heering and Falk Müller, "Cultures of Experimental Practice: An Approach in a Museum", *Science & Education* 11 (2002), pp. 203–214. One out of five sections of the exhibition addressed the salon science in the late eighteenth century. Electricity played a key role in this section.

3 A number of instruments had been initially reconstructed in research projects. Several of the instruments used in the exhibition were duplicates of these instruments, and were specifically made or purchased for the exhibition.

FIGURE 12.1   Three electrical 'toys'. From left to right: electrical boxer, electric hailstorm and electrical huntsman
© EUROPA-UNIVERSITÄT FLENSBURG

a spherical conductor. In the device used in the exhibition, the glass disc was replaced by a plastic one.[4] The second device was a thunder house, a device used to demonstrate the protective mechanism of a grounded lightning conductor. In our model house there were two chimneys; on one of them a wire ran down which could be grounded and thus demonstrated the effect of the lightning conductor. On the other chimney a wire led into a small tin box inside the house, which contained a spark gap. The house had four side walls, which were made of thin planks leaning against each other, and on which two thin planks serving as a roof were laid.

Before the demonstration began, an ethylene-air mixture was created in the can, which is highly explosive and could be ignited by the spark.[5] At the beginning of this part of the demonstration the participants were familiarised with the purpose of the house and warned of an imminent loud explosion. From a loaded Leyden jar, a clearly visible spark struck the grounded wire on one chimney as the demonstrator approached: nothing happened. Those present

---

4   This was done in order to meet safety regulations; a glass plate can break during the use of the machine.
5   Historically, gunpowder was ignited which was not allowed due to safety regulations. The thunder house follows a design that we learned from Ellen Kuhfeld (The Bakken, Minneapolis).

RE-ENACTING HISTORICAL EXPERIMENTAL PROCEDURES 229

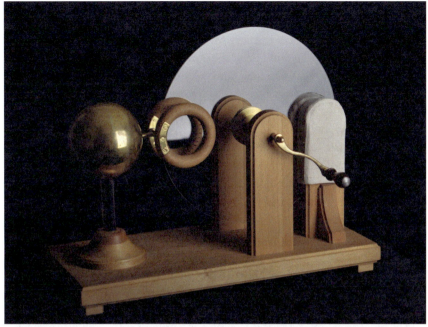

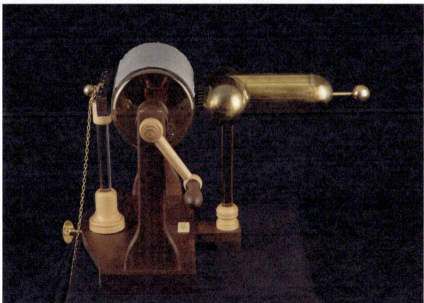

FIGURE 12.2   Two electrostatic generators
© EUROPA-UNIVERSITÄT FLENSBURG

were warned that the next discharge would hit the ungrounded (and thus dangerous) chimney. The button of the loaded Leyden jar was moved towards the chimney, a spark flashed over, there was a very loud detonation and a flash of fire, the house collapsed and after a short moment, pent-up tension and shock were discharged into laughter and conversation.

Most instruments were placed on a large wooden table, some instruments that were not supposed to be used were placed on shelves. A series of posters was hung on the walls, through which the historical, but also the physical context of the electric salon was illustrated as a theme. In addition, a print of the painting *An experiment on a Bird in the Air Pump* by Joseph Wright of Derby was hung on one of the walls. Next to it there were some (old-looking) chairs in the room, in an attempt to create an atmosphere that differed significantly from that of the other rooms in the Phänomenta.

Unlike the rest of the Phänomenta, this room was not accessible all the time but had special hours in which a demonstration was scheduled. The demonstration took about one hour, and, in addition, the participants had to book this event (initially with extra costs of one Euro per person; in the last year, participation was free).

The electric salon was introduced by a member of staff, who was also present during the whole time and guided the activities of the visitors. At the same time, this member of staff explained the historical context of the experiments, gave some information about the physical background and answered questions. Except for the thunder house which was demonstrated, participants were invited to interact with the instruments and to aim at creating the respective effects for themselves. The demonstrator gave some hints to the visitors; in addition, historical engravings served as instructions for the activities. During the electrical salon there were further activities that were instructed, as well as those that were left open to interpretation by visitors. Stronger instructions (verbally by the guide or by illustrations on the posters) were used for direction especially when new equipment that had not yet been introduced was to be used. During the introduction to the salon, the historical contextualisation showed that electricity had a completely different meaning at the beginning of the eighteenth century than in the twenty-first century. An electrified body can be defined simply by its ability to attract light fabrics. In the activity that followed the introduction, various rods (transparent glass, white glass, black glass, plastic, steel, wood) were rubbed with different materials (wool cloth, silk cloth, rabbit fur) and these rods were tested to see if they were able to attract light paper scraps. Similar experiments were then carried out, in which an insulated person was charged with an electric machine and tried to attract

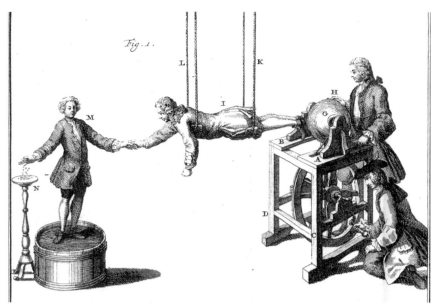

FIGURE 12.3   Plate with electrically charged person attracting light objects. From John Theophilus Desaguliers, *De Natuurkunde Uit Ondervindingen*, Isaak Tirion, Amsterdam, 1751, 3 vols.: v. 2, pl. 38
REPRODUCED WITH PERMISSION OF THE LANDESBIBLIOTHEK OLDENBURG

light bodies with his hand.[6] This was the first of several activities in which the human body was involved in the experience. It was essential that such activities were carried out, but it was equally essential that these activities started with subtle effects and were increased only gradually.[7]

Following these initial activities, others were carried out in which the human body was also central: small sparks were pulled out of fingers and people were electrically charged to experience what this feels like. All these activities exhibited an entertainment character, which was also relevant in the historical situation. Furthering that emphasis, electric toys were shown next. Then the Leyden jar was introduced and a discharge circuit was realised. For this purpose, all participants who wanted to do so formed a circular chain by holding hands.

---

6 See Fig. 12.3 for a plate used in the posters and Fig. 12.4 for the activity.
7 At the same time, it was also important that the fear that many people have of electricity is taken seriously and respected, but also that the unreflective fear of all electrical phenomena is reduced.

FIGURE 12.4   Electrically charged person attracting light objects
PHOTO BY C. ANRICH, PHÄNOMENTA FLENSBURG

One person at one end of the chain held the loaded Leyden jar in one hand and another person at the other end of the chain touched the button that was connected to the inside, thus bringing about the discharge that all the people in the circle could feel simultaneously. After everyone shared this electric sensation, the concept of the electric circuit was discussed and the activities were modified according to the participants' ideas; for example, sometimes parallel circuits of two partial parts could be realised with larger groups. In this participatory context, it should be emphasised that these experiences were not only addressed from a physical perspective, but also from a historical and social one. Thus, for example, eighteenth-century itinerant lecturers were discussed, but also the political implications of enlightened experimentation. An essential aspect here is the claim to understand the world and the processes in the world by means of one's own mind and not to rely on traditional authorities. But there are also experiments in which this becomes clearer. For example, at the court of Versailles in 1746, Jean-Antoine Nollet discharged a Leyden jar through 180 gendarmes as well as over 200 Carthusian monks.[8] In doing so, he made – literally and figuratively – the representatives of secular and ecclesiastical power jump,

---

8   See: John Heilbron, *Electricity in the 17th and 18th Centuries: A Study of Early Modern Physics*, University of California Press, Berkeley, 1979, p. 318.

thus demonstrating a claim to power of experimental natural research. In the electrical salon, subsequently, the electrophorus (at which the phenomenon of induction could be discussed) with Lichtenberg figures was also addressed. At the end of the electrical salon, first the thunder house was presented and then the crown of cups was investigated by the audience putting their fingers into the glasses. With only some elements included in the electrical circuit, there is nothing for their fingers to feel (except that the salt water feels wet and cold). But if a combination of the glasses is chosen in such a way that more and more elements are short-circuited, then at some point they will feel a prickle, sting or tingling sensation that increases, the more elements are added. The threshold is by no means the same for all people: small injuries, such as those caused by gardening, on one of the immersed fingers increase sensitivity enormously. On the other hand, people who play stringed instruments are very insensitive due to the resulting calluses on their fingertips. The intention behind the two final demonstrations is different. The thunder house demonstration focuses on the question of the usefulness of electricity and electrical research in the eighteenth-century perspective – an aspect highly relevant in the context of the Enlightenment. The crown of cups, on the contrary, created a parallel to the discussion at the very beginning of the salon. Once again, an unfamiliar phenomenon was examined by perceiving its effect on one's own body.

During the various activities of the electrical salon, we aimed to contextualise them. For example, we provided historical background and scientific information along with suggestions for manipulating the apparatus. As already mentioned, posters attached to the walls described specific aspects of the historical context. Electrical generators were a topic as well as measuring devices. The historical context included such practical applications as medical electricity and the lightning rod. The museum guide narrated from this history and developed an interaction with the participants, thus not telling a standard story, but addressing topics that were relevant for the individual participants. Having made reference to the posters, the guide invited guests to read these in detail at the end of the visit.

At the end of the demonstration, participants were invited to give feedback. In the first year, we used a questionnaire designed on a four choice Likert scale.[9] Typical choices were: "fully agree – partly agree – partly disagree – strongly disagree". Items addressed the overall experience as well as details such as: the texts on the posters; the information given by the demonstrator; and the

---

9   We also carried out short oral interviews with the staff that instructed the participants. However, it would go beyond the scope of this paper to discuss these results in detail.

interplay of different aspects such as the activities, the historical and the scientific background information. Completed questionnaires were evaluated by the IT support of the Europa-Universität Flensburg.[10]

As can be seen, the overall feedback was positive (Fig. 12.5),[11] however, one problem also became evident from the data analysis: the evaluation of the questionnaires also showed that many of the respondents gave the same answers for many items. This is very striking; for example, for item B (the objects, namely, the instruments, were interesting), between 50% and 75% of the respondents completely agreed with this statement. This makes a more differentiated analysis almost impossible: the respondents can of course be divided into different groups (for example, according to age, gender, whether the visit to the electrical salon was at the beginning or end of the survey period, and so on). Since the overall group shows a very homogeneous response behaviour, there is also no differentiation for subgroups composed according to criteria. Consequently, in the following year, we used an eleven choice Likert scale. To sketch but one example, when we asked about the satisfaction with the electrical salon, these were some of the choices:

100%   This exhibition is excellent, nothing can be improved
90 %   The exhibition is almost perfect, some minor aspects can be improved
80%   The exhibition is great, most of the time it was exciting
70%   The exhibition is super, one can learn a lot of interesting things
...   ...
20%   The exhibition is very poor, most of the time I was bored
10%   The exhibition is lousy, almost everything was dissatisfying
0%   The exhibition is a disaster, absolutely nothing is good

The diagram in Fig. 12.6 shows the agreement of all visitors of the electric salons, those of the people who had visited the Phänomenta anyway and also visited the electric salon (general), and those of the people who came to the Phänomenta especially for the visit of the salon (special).

---

10   It should be noted that the collected data are ordinal but not metric, i.e. a ranking of the valence can be created, but not a mean value, since the value "2" is not twice as large as the value "1". For this reason, the corresponding items are evaluated using box-plots.

11   Box-plots are a way to visually represent survey data that is not metric. For such an analysis, the answers are first 'sorted' according to their approval. The median separates the upper half of the consent from the lower half (at N = 114 the median is between the answers 57 and 58) and is marked with a black line. The box contains (at least) the 25% of answers above the median and 25% of answers below the median.

RE-ENACTING HISTORICAL EXPERIMENTAL PROCEDURES

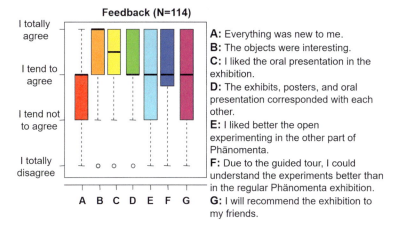

FIGURE 12.5   Boxplot for the evaluation of the electrical Salon, N = 114

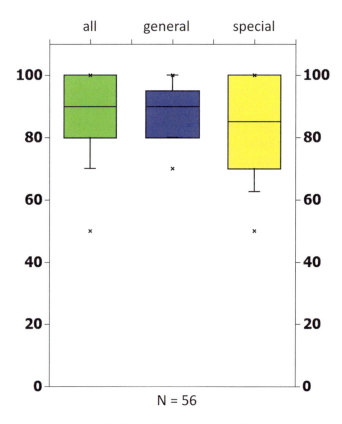

FIGURE 12.6   Boxplot for analysis of visitors' satisfaction, year two, N = 56

As one can see from the evaluation of all responses, we still got extremely positive feedback. However, and this is especially important with regard to the further analysis of the data, the diagram now also shows a greater spread of the answers. This spread now makes it possible to perform a more differentiated analysis. For example, the entire group can be divided into two or more subgroups according to their response to an item. These subgroups can then be statistically compared with regard to their response to other items. For example, a distinction can be made as to the reason for visiting the electric salon. The subgroup associated with one reason can then be compared with the subgroup that responded a different reason, to elucidate whether these two subgroups exhibited perceptions that differ from each other in statistically significant ways.[12]

Subsequently, we analysed two further aspects, for each of which we chose further methodological approaches to analysis. In a first study, we used semi-structured interviews to investigate why visitors to Phänomenta were not in the electrical salon. From a total of 93 interviews, it became clear that there were two main reasons for this behaviour: on the one hand, there were a number of visitors who, due to time constraints, feared they would miss too much of the rest of the Phänomenta during their visit; on the other hand, there were also a larger number of visitors who had not even noticed the electric salon. This was certainly at least partly due to the location of the production: the access to the North Gate is rather hidden, and in this respect the choice of location was certainly not an ideal one.[13]

In the second study, we used a questionnaire that visitors to the salon could fill out before leaving the room. Among other things, we used a ranking system to find out about the relevance of individual aspects for the visitors. We defined six items, each of which was to be assigned one of the ranks from 1 (most important) to 6 (least important). It was found that (significantly) the most important aspects were the personally-conducted demonstrations and the demonstration of experiments. However, it was also highly significant that the location (the room in the northern gate) was the least relevant for the positive impression.

Based on the combined results of these two evaluations, we decided to move the electrical salon from the beautiful room in the Nordertor to a very

---

12   This analysis is not based on interpreting the box-plot diagram, but uses statistical tests such as the Wilcoxon test.

13   For a more thorough discussion of this study, see: Peter Heering, Martin Panusch and Vanessa Schmid, "Evaluation der Sonderausstellung 'Elektrischer Salon' in der Phänomenta Flensburg", *PhyDid B* (2014), www.phydid.de/index.php/phydid-b/article/download/469/609, 1 (accessed 31 Oct. 2020).

technical and modern room in the middle of the Phänomenta where its visibility was significantly increased. Not surprisingly, this led to a significant increase of the number of visitors who booked this programme.

However, despite the increased number of visitors, there was still a problem: the electric salon was too expensive compared to many other Phänomenta event offers because of the necessity of having to have a member of staff in the room with the visitors. Since this activity remained a subsidised operation, it was finally decided to no longer offer the electric salon at Phänomenta. However, we are still receiving inquiries at the university – both from teachers who want to visit with school classes and from groups of adults. At the moment, such activities can still be offered at the university.

3 Enabling Secondary School Students to Make Their Own Instruments

When it comes to the implementation of historical instruments in formal education, a key problem is the availability of the instruments. Schools in the Flensburg neighbourhood can access or borrow instruments from our collection in order to implement these devices into lessons – as some former students are now teachers in these schools, they are aware of this option. Moreover, we have realised several outreach activities that addressed teachers in particular, in order to make them aware of the approach and of the opportunity to get supported. However, for schools that are not local to Flensburg, this gets too time consuming and is therefore not an option. As a result, from these difficulties, we developed a project (Projekt Galilei) that tried to overcome this challenge in a completely different manner. In doing so, we tried to use the opportunities that have recently been implemented in the German educational system.

A few years ago, schools were given the opportunity to introduce elective courses as part of the newly established all-day classes. These courses were open in content, with the exception that the topics of the compulsory lessons were excluded. This is where our offer came in. The project had two phases: in the first, we offered an on-campus teacher training; in the second, we supported the actual course that was taught in school classrooms. During the two-day advanced training course, teachers were able to gain an insight into the history of physics and its educational benefits. At the same time, they were introduced to a number of historical devices that we believed to be fundamentally suitable for reproduction in a school context (among them two *cameras obscura* with different dimensions, each with a lens and a focal unit; see Fig. 12.7).

FIGURE 12.7 Two *cameras obscura* used in Projekt Galilei
© EUROPA-UNIVERSITÄT FLENSBURG

FIGURE 12.8   Teacher working on an astatic galvanometer during the workshop
PHOTO BY M. ENGEL, EUROPA-UNIVERSITÄT FLENSBURG

The teachers also had the opportunity to build one of the devices themselves as part of the training. For this purpose, materials and tools were provided so that the teachers could successfully complete their apparatus in the short time available and then take it home (Fig. 12.8). In the pilot study, four teachers participated; two of them were able to enter the second phase.

In the second phase, these teachers offered an elective course in their school that was devoted to reconstructing a historical instrument. The concept for this course was to have a first general introduction, then introduce some instruments to the students by showing some images and explaining the technical aspects. Thus, the students were able to decide which device they wanted to work on. They were provided with a replica of the corresponding device for a few weeks. This replica was examined by the students with the goal of building their own version of this device, which would then remain in the school and be used in the lessons there. There were a couple of conceptual ideas connected with this approach. Students were supposed to understand that science is not only a cerebral activity, but also that artisanship has a relevant role in the process of generating scientific knowledge. At the same time, they were to be encouraged that potentially, they could make a career in science in the role of technician or artisan. In this respect, it was relevant that these courses not only appealed to students who were high-performing in the natural science subjects, but also to students who had lower grades and were about to leave secondary school after ten years of education. The aim of this course was to give these students the opportunity to achieve success in the context of physics and to experience their self-efficacy. In addition to these more individual-oriented goals, there was another objective: not only should replica historical instruments be part of the school collection, but with the knowledge that these instruments were made by students of this school, these instruments were given a different status. Since associates, who also created ownership of these instruments,[14] made them, these objects may be clearly distinguished from the usual instructional devices in the school collection, which come from the commercial teaching aid companies.

Two schools were involved in the pilot study. In one school, the students worked for one year and reconstructed three instruments: a *camera obscura* (with a lens), a demonstration of the superposition of motion (according to Musschenbroek), and a water prism (according to Goethe). In the second school, the students worked for five months and decided to build a *camera obscura* and an electrostatic generator.

The activities in the two schools were evaluated; both a questionnaire and semi-structured interviews were used.[15] The evaluations examined the students' perceptions of this approach, what they liked or disliked, and the skills they developed. From the analysis, it became evident that the students of the first school perceived the course significantly more positively than the ones of the second school. However, at the time of the survey, the students of the first school had already completed their first device, while the participants of the second course, which started much later, were still working on their first device. In this respect, it has to be questioned whether this result could possibly be interpreted as an artefact of the time of the survey. In addition, observations made by the teachers during the activities of the students showed that the students brought physical concepts into the planning and implementation of the devices, for example by determining appropriately what dimensions the *camera obscura* should be given in relation to the lens. More insights resulted

---

14   On the role of ownership, see in particular Don Metz and Art Stinner, "A Role for Historical Experiments: Capturing the Spirit of the Itinerant Lecturers of the 18th Century", *Science & Education* 16, 6 (2007), pp. 613–624.

15   The evaluation was carried out in: Sören Schubert, *Schülerperspektiven im 'Projekt Galilei': Wie Schüler der Sekundarstufe I über das selbstgesteuerte Lernen an historischen Experimentierstationen denken*, unpublished MA-thesis, Universität Flensburg, 2011.

RE-ENACTING HISTORICAL EXPERIMENTAL PROCEDURES   241

from semi-structured interviews with some of the students. For example, a female student from the course stated in an interview that she wanted to become a carpenter and had therefore already quickly grasped how to build the wooden housing of the *camera obscura*. Besides, it became clear in the interviews that it was not only the building of something that was perceived as interesting, but also the experimenting with the devices. Here it was especially the colours created with the water prism that fascinated the students.[16] It is certainly not surprising that phenomenological aspects in particular were appealing to the participating students – mathematical considerations were only made when necessary in the context of building the devices.[17]

4   Conclusion

Using historical instruments (or their reconstructions) provides unique learning opportunities. Implementing these instruments creates opportunities to develop an understanding about the various aspects that are relevant in experimental work. The two approaches presented here have in common that the interaction with the instruments leads to an understanding that is not limited to the physical aspects. Rather, the respective contextualisations of the experiments in the context of the electric salon are essential. In this approach, experimental practice can be experienced as culturally shaped and culturally relevant. At the same time, it becomes clear that the production of physical knowledge is by no means a linear success story, but that it also leads to problems of interpretation, reinterpretation and dead ends. In addition, it also becomes clear that scientific research is by no means exclusively a cerebral activity, but that both the production of instruments and their use require corresponding manual skills. Consequently, these approaches are not only relevant for education in science, but also in understanding the nature of science.[18]

---

16   *Ibid.*, "Appendix A".
17   For a more detailed discussion of this project, see: P. Heering, "Make – Keep – Use: Bringing Historical Instruments into the Classroom", *Interchange* 46, 1 (2015), pp. 5–18.
18   On teaching Nature of Science in general, see for example: Douglas Allchin, *Teaching the Nature of Science: Perspectives & Resources*, Ships Education Press, Saint Paul, 2013; William F. McComas, *The Nature of Science in Science Education: Rationales and Strategies*, Kluwer Academic Publishers, Dordrecht-Boston, 1998; William F. McComas (ed.), *Nature of Science in Science Instruction. Rationales and Strategies*, Springer, Cham, 2020. On the use of historical experiments in this respect, see in particular: P. Heering and E. Cavicchi, "Teaching About Nature of Science Through Historical Experiments", in *Ibid.*, pp. 609–626.

In this context, it seems essential that the assessment of these educational offerings is not based solely on anecdotal assessments based on individual statements made by participants. Rather, it seems essential that empirical evidence is also created that, in conjunction with the statements made, provides a much more differentiated but also more robust picture of the effects of these offerings. It should be noted, however, that the empirical findings presented here are by no means sufficient to create reliable statements on these activities. Rather, they are to be understood more in the sense of pilot studies which, on the one hand, show that such research promises to produce relevant results. On the other hand, they allow the hypothesis that such uses of historical instruments (or their replicas) create extremely valuable options for educational processes.

Acknowledgements

The electrical salon was realised with the support of Phänomenta. Particular thanks go to Martin Panusch who substantially supported the evaluation in this project. This study also benefitted from the independent research of Vanessa Schmid. Projekt Galilei was financially supported by the Nordmetall-Stiftung. The project was developed in collaboration with Friedhelm Sauer, and supported by Martin Engel and Martin Panusch.

CHAPTER 13

The Lorentz Lab: Reviving the Scientific History of Teylers Museum with Working Replicas

Trienke M. van der Spek | ORCID: 0000-0001-6163-6043

1 Introduction

In 2017, Teylers Museum opened the *Lorentz Lab*, a new extension to the permanent presentation of the museum. It is the first tangible result of a long term programme taken up by the museum, aiming to revive Teylers Museum's long history as an institute of art and science, and to make this context more accessible to the public without disturbing the historic museum presentations. The Lorentz Lab is specifically dedicated to the institute's scientific past, starring one of the least communicative collections in the museum: the scientific instruments. The Lorentz Lab brings the unique origin of this heritage into life and dialogue with the audiences of Teylers Museum in a carefully recreated historical setting of Teylers' Physics Laboratory, where working replicas and theatrical support allow visitors to participate in science directly and engage with the activities and scientists that shaped Teylers Museum and its collections. The development of the Lorentz Lab, with its challenges, experiences and insights from an instruments' point of view is central to this chapter.[1]

2 Teylers History on Display

Teylers Museum is the oldest museum of The Netherlands and opened in 1784 as a laboratory and workplace for science and art; a place where ordinary citizens could ponder on the great questions of their day, nourished by the knowledge and learning to be found in books, fossils, instruments and works of art. Although immediately named 'museum', the focus was contemporary and

---

1 This chapter is based on my presentation of the Lorentz Lab at the *XXXVII Symposium of the Scientific Instrument Commission*, Leiden and Haarlem, The Netherlands, 3–7 September 2018.

© TRIENKE M. VAN DER SPEK, 2022 | DOI:10.1163/9789004499676_015

active, in line with the eighteenth-century definition of a 'mus*ae*um'; a place of modern knowledge, inspiration and contemplation.[2]

Only over time was Teylers Museum transformed into a museum of its own rich past, and the original eighteenth and nineteenth century museum rooms form an inseparable part of it.[3] The buildings, collections, archives and historical displays of Teylers Museum cover a span of nearly 250 years and were all originally bought or created in the service of contemporary activities in science and art. The entirety of this site survived time in its full original context and as such, is internationally valued as unique heritage.[4]

Teylers Museum hopes to give this background a more prominent place in the museum's public outreach with a long term program. Besides the Lorentz Lab, this programme will include a broad range of stories that highlight different aspects of Teylers' institutional history. Examples are Napoleon's audiotour, based on his visit in 1811 that shows visitors the museum and its early nineteenth-century settings through his eyes, and *Pieter Teyler's House*, opening at the end of 2021 in the restored former living quarters of Pieter Teyler van der Hulst (1702–1778) – also the seat of Teylers Foundation since 1778 – and dedicated to bringing to life his personal story, his ideals and his testament, which led to the foundation of Teylers Museum.

3 Science at Teylers Museum

Upon opening, Teylers Museum addressed a broad range of scientific topics that was fashionable in many eighteenth-century learned societies, covering areas of natural history as well as the exact sciences. This broad scope remained well into the twentieth century and is still represented by the collections of fossils, minerals, instruments and richly illustrated botanical works in the historical library of the institute.

---

2 See: Martin P.M. Weiss, *The Masses and the Muses: A History of Teylers Museum in the Nineteenth Century*, PhD dissertation, University of Leiden, 2013, pp. 85–89.

3 The interest in and display of the institute's own history already started in the nineteenth century. See for the instrument collection: M.P.M. Weiss, "'Monuments of Science': how the Teyler Museum's instrument collection became historical", in Jim Bennett and Sofia Talas (eds.), *Cabinets of Experimental Philosophy in Eighteenth-Century Europe*, Brill, Leiden, 2013, pp. 195–213.

4 Including two learned societies and the original foundation executing Pieter Teyler's testament, all dating from 1778, and an eighteenth-century social housing complex (Teylers Hofje) close to the museum complex.

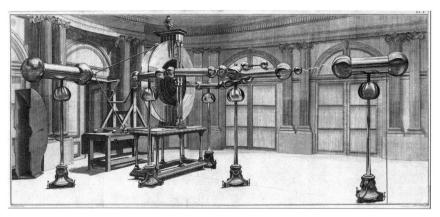

FIGURE 13.1　The large electrostatic generator producing long sparks in the Oval Room of the museum, as illustrated in Martinus van Marum, *Beschryving eener ongemeen groote Elektrizeer-Machine, geplaatst in Teyler's Museum te Haarlem*..., Enschedé & Walré, Haarlem, 1785, pl. 1

By contrast the Lorentz Lab focuses specifically on Teylers' activities in the exact sciences. Research in this area immediately gained weight upon opening of the museum in 1784, thanks to the first director of Teylers Museum and man with a mission, Martinus van Marum (1750–1837). His *Ongemeen Groote Elektrizeer-Machine* (uncommonly large electrostatic generator) in the Oval Room (Fig. 13.1) and the experiments performed with it in the museum's first years immediately attracted international scholarly interest and quickly established Teylers Museum as a scientific institute of importance.[5] Like van Marum, his successors in the nineteenth and first half of the twentieth century were fully paid scientists commissioned to do research. Physicist and Nobel Prize laureate Hendrik Antoon Lorentz (1853–1928) – who directed Teylers Physics Research Laboratory between 1909 and 1928 – is its most prominent representative, and the Lorentz Lab, located in this former physics laboratory, has been named after him.

As part of this scientific enterprise, new instruments were constantly acquired for research and demonstration purposes. Over time, additional museum rooms were added to the Oval Room to house these growing

---

5　For more on how establishing the importance of Teylers was part of the agenda and explicitly aimed at, see: Gerhard Wiesenfeldt, "Politische Ikonographie von Wissenschaft: Die Abbildung von Teylers 'ungemein großer' Elektrisiermaschine 1785/87", *NTM International Journal of History and Ethics of Natural Sciences, Technology and Medicine* 10 (2002), pp. 222–233.

FIGURE 13.2   The 1885 instrument room with van Marum's centrally placed large electrostatic generator and Leyden jars, surrounded by cabinets filled with scientific instruments
PHOTO TEYLERS MUSEUM

collections, and dedicated science laboratories were built to meet contemporary requirements and specialisations in scientific research.[6]

The accumulated heritage of this scientific past is only partly visible in Teylers Museum today. Two museum rooms – including the famous eighteenth-century Oval Room – house the instrument collection (Fig. 13.2) and are publicly accessible, but others like the remaining laboratory buildings

---

6   A more detailed account on the scientific activities of Teylers as an institute can be read in: Weiss, *The Masses and the Muses…*, cit. (n. 2); Gerard L'E. Turner, *Descriptive Catalogue of van Marum's Scientific Instruments in Teyler's Museum*, «Martinus van Marum. Life and Work 4», Noordhoff international Pub., Haarlem, 1973; G.L'E. Turner, *The Practice of Science in the Nineteenth Century: Teaching and Research Apparatus in the Teyler Museum*, Teylers Museum, Haarlem, 1996. Additional information can also be found in biographical studies on van Marum and Lorentz. Not the most recent, but most comprehensive and in English, study on van Marum is: R.J. Forbes (ed.), *Martinus van Marum. Life and Work*, Tjeenk Willink, Haarlem, 1969–1976, 6 vols. On Lorentz see for example: Abel Streefland, *Hendrik Antoon Lorentz, de Haarlemse jaren van een wereldberoemd wetenschapper, 1909–1928*, «Teylers Museum Magazijn 131», Teylers Museum, Haarlem, 2017; Frits Berends and Dirk van Delft, *Lorentz, gevierd fysicus, geboren verzoener*, Prometheus, Amsterdam, 2019; Anne J. Kox, *Hendrik Antoon Lorentz, natuurkundige 1853–1928*, Balans, Amsterdam, 2019.

have been inaccessible to museum visitors until recently.[7] The most invisible part of the scientific heritage, however, is the original context of science dynamics that created the museum buildings and collections. The instruments on display have always been difficult for the general public to understand,[8] but as historical objects their original purpose and meaning is even less discernible to modern day museum visitors. The Lorentz Lab tries to bridge this gap by re-creating the historical science dynamics with appealing educational activities.

4 The Lorentz Lab

The Lorentz Lab brings these dynamics of the past to life and connects it to the instrument collection with a mix of theatre, evocation and working replicas of instruments. In a non-formal learning environment two different target groups are addressed: general visitors and school groups. Two separate programmes have been developed: *The Lorentz Formula*, for the general public and *Einstein was here* for school groups.[9] Each programme has a different emphasis: *The Lorentz Formula* focuses on Lorentz personally and highlights his contributions to science, society and his position at Teylers; *Einstein was here* combines high level physics experiments with the history of discoveries in electricity. The programmes share the goal of taking visitors back in time and making them participants in the scientific activity that was once central to Teylers Museum.

The Lorentz Formula is a location theatre performance enacted by professional actors, and presented to general audiences.[10] This drama takes visitors on a voyage through time. Setting off from the time of Lorentz, the audience and actors travel back to the start of the research tradition at Teylers Museum in the eighteenth century. The working replica of the large electrostatic generator of 1784 provides the thunderous finale. During this performance, the visitors participate in experiments and think about the great questions and discoveries of historical physics, such as Einstein's theory of relativity. The result is a very accessible immersive public experience that brings people,

---

7   Except incidental tours. The small Garden Observatorium of 1866 is used for educational programmes.

8   Elisa van der Ven, *Gids door de verzameling Physische Instrumenten*, Erven Loosjes, Haarlem, 1898, p. 18. Van der Ven e.g. describes a demonstration table in the oval room with modern electrical equipment, demonstrated by the museum guide upon request.

9   *The Lorentz Formula* runs twenty times a week with a capacity of twenty people at a time. *Einstein was here* is booked on request.

10 Creative development and production of The Lorentz Formula by Rieks Swarte and De Toneelschuur (Haarlem Municipal Theatre), in joint venture with Teylers Museum.

ideas and instruments to life. Visitors feel that they have really learned something fundamental about the study of physics, and sometimes wish they had done so earlier – or as one visitor framed it "if only I'd had physics from you [the actors] at school, I would have been become much wiser and happier ..."[11]

By contrast with that theatre drama oriented to the general public, *Einstein was here* is a high-level educational programme, offered only to physics classes by advance arrangement with teacher and school. *Einstein was here* provides a two hour programme on location, preceded by a preparatory lesson at school and concluded with lessons afterwards to process and discuss the interconnected results in the classroom. In the Lorentz Lab, pupils work for one hour on nine different experimental set-ups, which will be described in closer detail in *Section 6*. In the additional hour they witness a demonstration of the big electrostatic generator replica and visit the original instruments in the museum galleries, where they discuss the histories of these objects with a trained guide.

This programme is designed to provoke questions that fit in with general educational requirements and to make links to the school curriculum specific to each visiting class. Participating pupils already have an exact sciences profile, and are in their last two years of secondary school.[12] Developed in close collaboration with physics teachers, this programme puts pupils to work actively and creatively in the Lorentz Lab.[13]

During their *Einstein was here* experience, pupils conduct fundamental research by carrying out experiments with working replicas of historic electrical research instruments. By doing experiments, they learn how humanity came to understand electricity and discover the origins of its modern application. The Lorentz Lab setting invites the pupils to be in the same historical time as imaginary colleagues of famous nineteenth-century scientists. Pupils actively participate in the original research quest and contribute to the historical discoveries with their own experiments in the Lorentz Lab, rather than just repeating or reflecting upon historical accounts.

4.1 Historical Embedding

The building itself plays an important role in the storytelling of both programmes: the Lorentz Lab is housed in the former physics research laboratory,

---

11   Responses from questionnaires, visitors' books and on social media.

12   *Havo* (school for higher general secondary education) 4th and 5th grade; *vwo* (pre-university education) 5th and 6th grade.

13   A group of physics teachers from nearby schools and feedback groups: workshops held in the museum and at the Physics Teachers Conference in 2014.

FIGURE 13.3   The Lorentz Lab: Furniture has been based on designs from a Max Kohl physics sales catalogue contemporary with the chosen historical era. Central left is a De La Rive Tube replica and original Ruhmkorff coil
PHOTO TEYLERS MUSEUM

dating back to 1791,[14] and last modernised under the direction of Lorentz in the 1910s.[15] After the closure of the laboratory, the building served different purposes before its restoration and transformation to a public room began in 2014. At that time, virtually nothing of the original interior was left, which gave the freedom to refurbish the laboratory in the service of the above-stated public goal.

A look and feel of the early twentieth century was chosen, taking visitors back to the moment Lorentz took charge of the laboratory (Fig. 13.3). For example, the presentation of Lorentz's study room suggests his activity and work are still in progress, with notes, letters and open books on his desk. The original functionality of the rooms has been kept as close to the original as possible and traces of the original colour scheme were incorporated in the re-styling

---

14   Van Marum opened the first Dutch research Laboratory of Physics in 1791 on this site. The current building dates from 1884. In 1955 research ended when the laboratory officially closed down.
15   Weiss, *The Masses and the Muses...*, cit. (n. 2), pp. 276–277.

of the laboratory wherever possible.[16] The evocative interior design has been implemented in the smallest detail – even the lift, toilets and all lighting are in contemporary style – making the building function as a multi-sensory time capsule that places visitors one hundred years back in time before starting their programme, in which again this historical embedding is all-important.

4.2 Working Replicas

Working replicas are an essential part of the Lorentz Lab, for demonstration purposes and to be operated by the public. Using them is crucial to the visitors' experience, and alongside the historical decor of the Lorentz Lab, these replicas are part of the time-machine that brings back the original action: in the historical laboratory, ready to be used in contemporary ongoing research, instruments from the collection take back their original identity as scientific tools. This is underlined by the deliberate choice not to replicate the historical patina on the brass, wood, and so on, but to present them as new, emulating the original appearance when acquired by Teylers Museum.

In the Lorentz Lab the working replicas can be found in two locations: in the laboratory on the first floor nine experimental set-ups, fundamental to the school programme, are placed on the laboratory benches. The nearly life-size replica (Fig. 13.4) of Van Marum's large electrostatic generator is the *show piece* and situated in a dedicated room on the ground floor.[17] This replica has an essential role in both programmes that are held in the Lab (*The Lorentz Formula* and *Einstein was Here*). It is also used for separate demonstrations to visitors of Teylers Museum.

Electricity is the central theme of all working replicas at Teylers Museum. This was a key field of interest in international physics throughout the long span of time in which Teylers Museum functioned as a research institute; from static electricity research in the eighteenth century to electromagnetism and electrons in the nineteenth and early twentieth centuries. Today's instrument collection still reflects these important developments. Electricity also connects two of Teylers' most famous scientists: Van Marum and Lorentz. Electricity has become an essential commodity in today's technological society. In the future, this role is only expected to grow, making it the integrative theme that connects past and present in the Lorentz Lab.

---

16   Anne Marie ten Cate, *Teylers Laboratoriumgebouw, bouwgeschiedenis en huidige toestand*, Ten Cate en Van der Wielen, Haarlem, 2014, pp. 47–57.

17   The replica is 90% of the size of the original, defined by the dimensions of the room in the Lorentz Lab, which could not be changed due to the monumental status of the building. The eighteenth-century original is 3.5 m high and over 7 m long.

FIGURE 13.4 The working replica of van Marum's machine. The manual operation is on the right and outside the cage
PHOTO TEYLERS MUSEUM

Though an obvious choice, electricity posed complex issues for working replicas. User safety was a big challenge and it took many creative solutions and bypasses to produce the Lorentz Lab working replicas. For example, mercury had to be replaced and electromagnetic radiation shielded, without disrupting the historical experience of visitors.

5   Van Marum's "*ongemeen groote electrizeer-machine*"

Teylers Museum's large electrostatic generator was designed and built by van Marum and instrument maker John Cuthbertson, and first used in December 1784 in the Oval Room. With the double glass plates of 165 cm in diameter the original instrument was, and still is today, the biggest electrostatic plate machine in the world. It was used intensively in the early years of the museum's existence. Quickly renowned for its role in science as well as for its impressive dimensions, it became an icon of Teylers own rich history in the nineteenth century,[18] prominently placed in the new instrument room of 1885 where it still stands today.

---

18   See Weiss, "'Monuments of Science' …", cit. (n. 3), pp. 198–200.

Despite its fame, the electrostatic generator's reliability was rather poor and van Marum's complaints were ample. With Cuthbertson he continued tweaking and improving the machine and his publications give detailed accounts of the machine's construction, its flaws and efforts to solve them.[19] An important challenge was the direction of the immense charges to the relevant parts of the machine, without incurring unwanted discharges of all sorts that reduced the machine's performance. Many alterations to the electrostatic generator were made to solve this issue. Dutch weather was another serious problem. In Haarlem the climate was mostly too damp to operate a friction machine and, to make things worse, the Oval Room had no heating. Van Marum describes how he would open all windows in the Oval Room in winter to let in frosty, dry air, and place heated sandbags against the isolating glass posts of the generator, to let the air humidity drop as low as possible. Under such circumstances the machine reportedly gave sparks as long as 60 centimetres.

5.1   *Authenticity versus Functionality*
Teylers Museum wanted the working replica in the Lorentz Lab to be as authentic as possible, both in looks and operation, but also be safe, reliable, easy to use, and with a high spark performance. To meet these requirements, many adaptations of the original design were necessary. Some were inherent to the very delicate construction of the original as, for example, the segmented, loose fitting glass supports that carry the plates and central axis in the original instrument. Official health regulations for visitors and collaborators mandated protection against loud noise, strong electromagnetic discharges and exposure to mercury. Further efforts were involved to tackle the original eighteenth-century issues of air humidity and the machine's uncontrolled discharging.[20]

---

19   M. van Marum, *Beschryving eener ongemeen groote Elektrizeer-Machine, geplaatst in Teyler's Museum te Haarlem, en van de proefneemingen met de zelve in 't werk gesteld,* «Verhandelingen uitgegeven door Teyler's Tweede Genootschap 3», Enschedé & Walré, Haarlem, 1785; van Marum, *Eerste vervolg der Proefneemingen gedaan met Teyler's Electrizeer-Machine in 't werk gesteld,* «Verhandelingen uitgegeven door Teyler's Tweede Genootschap 4», Enschedé & Walré, Haarlem, 1787; van Marum, *Tweede vervolg der proefneemingen gedaan met Teyler's Electrizeer-Machine,* «Verhandelingen uitgegeven door Teyler's Tweede Genootschap 9», Enschedé & Walré, Haarlem, 1795.

20   The technical development and subsequent improvements leading to the current working replica has been discussed in detail in: Wolfgang Engels, "The Reconstruction and Rebuilding of Van Marum's 'Ongemeen Groote Electrizeer-Machine' and Some Experimental Results", paper presented at the *XXXVII Symposium of the Scientific Instrument Commission*, Leiden and Haarlem, The Netherlands, 3–7 September 2018; T. van der Spek, "Een 'ongemeen groote electrizeermachine' komt tot leven", *Nederlands*

THE LORENTZ LAB 253

The resulting instrument looks like an identical copy of the original electrostatic generator, with hand cut woodcarvings, nearly identical mahogany, and mouth blown glass.[21] Nonetheless, a lot of modern technology is hidden underneath to meet the standards described above, mostly invisible, for example stainless steel components and ball bearings, or carefully integrated into the historical appearance of the replica, like the dual operation system that enables an easy switch between manual and electrical operation of the machine.[22] Despite these modern adaptations, the replica retained the full functionality of a plate friction machine. With the possibility to use it manually, the overall performance as well as the visual presentation stayed surprisingly close to the original. The Faraday cage surrounding the working replica has been a necessity to protect audiences and personnel against electromagnetic fields and discharges. Although first seen as a disturbing anachronism, it is now used as an asset: the cage is part of the storylines in the Lorentz Lab. The Faraday cage is discussed with the audiences as an instance of an important scientific discovery that fits within the central theme of electricity, while at the same time it operates in a functional setting with the electrostatic generator operating inside it.

The emphasis on authenticity in looks and operation in combination with the other demands came with a price: the development of the replica was difficult and costly. And because the functionality as a true friction plate machine was kept, the machine needs labour-intensive maintenance to keep dust and abrasion on the instrument's disks and rubbing cushions under control. But it is a price that has paid back, because the working replica has become a public favourite over the past few years. It is highly valued by visitors as an authentic encounter with Teylers' institutional history and an exciting experience. Initially only demonstrated at the end of *The Lorentz Formula* as a theatrical apotheosis and as part of the school programme *Einstein was here*, the public demand was such that separate demonstrations are now being held during weekends and holidays.

Tijdschrift voor Natuurkunde 83, 5 (2017), pp. 146–149. A shorter, English version was published online:

"An 'Uncommonly large Electrostatic Machine' Comes to Life", *Dutch Journal of Physics* 2 (2017), www.ntvn.nl/magazines/2017-2/An%20uncommonly%20large%20electrostatic%20machine%20comes%20alive_1.html (accessed 6 Apr. 2021).

21 The replica was made by HistEx GmbH, Oldenburg in cooperation with Teylers Museum. The original mahogany species is on the red list; a sustainable alternative was chosen to replace it.

22 A dual operating system has been incorporated in the replica: with a smart switch the machine works with an electric drive or with a manual system that is equal to the original instrument.

6 The Smaller Replicas

As with the big replica of van Marum's electrostatic generator, Teylers Museum wanted the smaller replicas (Figs. 13.5 and 13.6) to stay close to their originals in look and operation while also ensuring safety and user requirements. In order for the public to work with electricity safely and directly, parts of the original instrumental setting had to be changed. For instance, no polluting dichromate batteries (Grenet cells) are used. Instead, all Lorentz Lab replicas work on a tailor made low voltage power supply, hidden in the middle of the tables.[23] This way the original features of the instruments could be kept intact, with their ample bare metal parts, by keeping currents and voltages within safe limits. The risk of small electrical shocks during operation – for which the trained guide warns pupils in advance – is accepted as part of the authentic historical experience. Shocks are part of the excitement that pupils experience, triggering curiosity and underlining the differences with modern day operation. Potentially dangerous parts have been carefully isolated though, like the top of

FIGURE 13.5   Working replicas. *Clockwise*: Spiral by Riess, with exchangeable tops with different windings; Ritchie motor; Clarke dynamo with different coils; Faraday's ring
PHOTOS ANTON STOELWINDER

---

23   Small replicas were made by Anton Stoelwinder, Gorredijk, with support of Jacob Slikker; the power supplies were developed by Ewie de Kuijper and René Overgauw.

FIGURE 13.6   Working replicas. *Clockwise*: Sine galvanometer; Ampère's Law Instrument; Oersted demonstration instrument; Voltaic pile. All replicas were produced in twofold
PHOTOS ANTON STOELWINDER

the De La Rive tube (visible in both Figs. 13.3 and 13.8), which is fitted with an isolating cap and warning icon.

Mercury was used routinely in nineteenth-century electricity experiments. Due to toxicity, this material is now banned. In the Lorentz Lab, this conductive liquid has been replaced by a modern and safe analogue, Galinstan. With this liquid provided in the contacts and troughs that were once filled with mercury, the replicas can still be used in their original operational manner (Fig. 13.7). This substitution allows pupils and teachers to actively get in touch with the important role that mercury played in the history of science. The characteristics of Galinstan bear a strong resemblance to mercury and are mostly unknown and exciting; very few pupils have seen or used real mercury at the present time. Using this 'liquid replica' gives this part of history a natural place in the Lorentz Lab experience and discussions, and illustrates how working with replicas in a close-to-real setting brings history alive and close-by.

FIGURE 13.7   Barlow's wheel replica, using Galinstan as mercury replacement in the trough. The replica is connected to the low-voltage power supply integrated in the laboratory bench
PHOTO TEYLERS MUSEUM

A few of the Lorentz Lab instruments are original historical objects. For example, the Holtz induction machine and the Ruhmkorff coil are genuine antiques that were acquired and refurbished for this purpose: these instruments are readily available on the market, usable and more affordable than building a replica of the same instrument. Pupils in the Lorentz Lab are impressed by the fact that these instruments are really old and that they are allowed to use them.

Finally, some alterations to the replicas are not part of the original, but facilitate the pupils' experimental quest in the Lorentz Lab. For instance, the Clarke dynamo (Fig. 13.5, bottom right) has two interchangeable coils with different windings, allowing pupils to examine and explain the difference in output between the two.

6.1   *Instruments at Work*

The replicas placed on the laboratory benches on the first floor of the Lorentz Lab are specifically selected for the teaching programme *Einstein was here*, a high-level physics educational programme for secondary school pupils with an exact sciences specialism. Pupils with this profile are in the last two years of their course and typically 15–18 years old. This teaching programme has been

THE LORENTZ LAB   257

jointly developed with physics teachers; from the choice of replicas to the didactic approach and physics course contents. Also, during the development, tests and workshops have been held with additional groups of physics teachers and pupils to finely tune the programme with their feedback. *Einstein was here* puts a strong emphasis on research skills and is based on the inquiry-based learning principle,[24] impressing the pupils with the importance of asking questions, of doing research and observing, and describing and interpreting the results.

Here pupils carry out experiments with working replicas of historic electrical research instruments in an open setting. They work in pairs and study a given phenomenon by designing their own working hypothesis and experimental set-up with working replicas and collecting experimental results. The pupils are addressed in a present tense, as if collaborating directly in association with historic scientists like Michael Faraday and Alessandro Volta. Pupils and historical scientists are teammates in the process of making historic discoveries. After set times, students who need it will receive a little help (tips) to ensure that everybody finishes in the one hour available with some results to process later. The supporting programme is provided on tablets to the pupils, which are also used to record the experimental results. These tablets connect to a specially designed e-learning environment that enables the pupils to report their findings and to discuss them together in the classroom.

There are nine experimental set-ups in the Lorentz laboratory covering major topics in electrical research spanning the nineteenth century. Each of these set-ups has one central instrument, as listed below. In addition, each set-up is outfitted with auxiliary equipment (for example, an electrophorus with the Holtz machine), measuring devices, connectors, and so on, all in historical design. The level of difficulty, among the experiments offered, slightly varies and allows teachers to match experiments with the capacities of their pupils in advance.

The nine central working replicas and related research themes are:
- Holtz machine (see Fig. 13.8): study of static electricity and influence. Also using an electrophorus.
- Voltaic pile (Fig. 13.6, bottom left): electric current, properties and principles of galvanic electricity.
- Ampère's law instrument (Fig. 13.6, top right): magnetic interaction between electric currents. Also using an Oersted demonstration instrument (Fig. 13.6, bottom right).

---

24   Translation of the experiments to the inquiry-based method and implementation in the Lorentz Lab by Technolab, Leiden, a company specialising in science education.

FIGURE 13.8   School group experimenting at the Lorentz Lab. On the left, the original nineteenth-century Holtz machine is being operated
PHOTO TEYLERS MUSEUM

– Faraday's ring (Fig. 13.5, bottom left): induction as a phenomenon.
– Spiral by Riess (Fig. 13.5, top left): using and quantifying induction.
– Ritchie motor (Fig. 13.5, top right): producing motion from current, identifying the efficiency.
– Clarke's dynamo (Fig. 13.5, bottom right): producing current from motion, defining electrical power.
– Sine galvanometer (Fig. 13.6, top left): electromagnetic relations and quantification.
– De La Rive tube (see Fig. 13.3): electromagnetic discharges, Lorentz force. Also using a Ruhmkorff coil and Barlow's wheel (see Fig. 13.7).

The historical setting is all-important and has been stressed by physics teachers as an important 'extra' to legitimise a visit to the museum in preference to doing a standard physics practical at school. Upon their suggestions, the programme is designed to reinforce awareness of how knowledge originates and grows. By involving learners actively with the historical, experimental and theoretical processes by which scientific knowledge is developed, this programme allies with "Nature of Science" (NOS), an international, research-based educational position. The NOS perspective views science instruction that is limited to facts and findings as inadequate; instead, science education for an informed

THE LORENTZ LAB   259

citizenry requires critical thinking and awareness of how humans make scientific knowledge, including the roles of evidence, subjectivity, controversy, observation, evolution in understanding, and diversity in participation.[25] By experimenting in company with scientists from the past, pupils learn how humans came to understand electricity; in doing so, pupils discover the origins of its modern application and form their own understanding of the relations among these discoveries and the underlying physics. Pupils appreciate this approach, and their opinion of the past develops during the programme, as illustrated by some quotes from two boys working with the Ritchie motor. At first mocking the replicas as "funny old machines", after one hour of using these machines, the boys' attitude changed. They exclaimed "how incredibly smart it is that people in the nineteenth century invented this way of producing motion with electricity!" Their visual judgement changed to appreciation of the historical instrument designers: "also making the machine look this beautiful too!"[26]

The uncompromising old-fashioned technology of the working replicas is an essential part of the students' personal encounter with the scientific enterprise of the past. In contrast to modern apparatus, a real understanding of the underlying physics principles is necessary to use these historical replicas successfully. A good example of a demanding replica in the Lorentz Lab is the sine galvanometer. Pupils use modern galvanometers at school, but this instrument is profoundly different: not a black box producing digits, but a mechanical instrument that functions by using the relationship between electricity and magnetism to measure currents. It is a fragile setting, asking patience and skill of the students and deductive insights in how the instrument works, so as to interpret their readings accurately.

This can be demanding and to keep pupils within time and within their comfort zone, this challenge has been focused on the central replicas in the experimental set-ups. Based on tests with pupils and feedback from teachers during the development of the programme, it was for instance decided to provide the pupils with familiar, modern measurement devices to use with the replicas. This also ensured reliable output to be used in their reports and discussions conducted afterwards at school.

---

25   Position Statement: Nature of Science, National Science Teaching Association, www.nsta .org/nstas-official-positions/nature-science (accessed 30 Mar. 2021); William McComas and Michael Clough, "Nature of Science in Science Instruction: Meaning, Advocacy, Rationales, and Recommendations", in William McComas (ed.), *Nature of Science in Science Instruction: Rationales and Strategies*, Springer, Cham, Switzerland, 2020, pp. 3–22.

26   Coen van der Kamp, "Antiek om mee te spelen", *NVOX* 43, 2 (2018), pp. 96–97: p. 96.

7   Experiences and Follow-Up

Three years after opening the Lorentz Lab, over 40,000 visitors have partici-pated in *The Lorentz Formula* and many more have experienced the working replica of Van Marum's big electrostatic generator. Critics in the world of thea-tre and science communication have been very positive. The Lorentz Lab was nominated for the European Museum of the Year Award 2020.

The museum aims to prolong *The Lorentz Formula* for another two years. In answer to school teachers' requests, it will be offered to school groups visits. Separate demonstrations with the big replica of the electrical machine have become very popular with visitors of Teylers Museum and will continue. More visitors will be offered personal interactivity with manual operation of the instrument.[27]

Einstein was here has been operational for the past two years with positive feedback from teachers, who praise the involvement of pupils, the attrac-tiveness of the experiments and the overall historic experience.[28] It also has successful derivatives, including a tailor-made programme for first year uni-versity students in physics,[29] and partial integration in another educational programme of the museum.[30] A recent evaluation of *Einstein was here* made clear that the original programme also has room for improvement.[31] A few logistical issues are outside the scope of this article, but two insights are worth discussing in closer detail. Both are related to the decision that was made for a very open, inquiry-based approach in the teaching programme. This popular didactical principle was highly valued by teachers during the developmental stage of the Lorentz Lab and experts in science education that were respon-sible for the educational approach of the experiments. But it requires pupils that are used to this independent way of working, and this is where theory may have met practice.

---

27   Bringing reliability and performance of the instrument up to standards first, implementa-tion of manual operation was slowed down.

28   Feedback from individual teachers after visiting the Lorentz Lab; see also C. van der Kamp, *Op. cit.* (n. 26).

29   Developed with the Technical University Twente, also available to other universities.

30   *Wow factor* for pupils of *vmbo* (lower vocational education): they do different experi-ments in the small garden observatory of the museum, and additionally visit the Lorentz Lab and discuss physics phenomena based on demonstrations of the large electrostatic generator and some of the small replicas.

31   WND Conference 2019 (annual conference for physics teachers), Noordwijkerhout, 13–14 December 2019, workshop setting and questionnaires.

Although a substantial group of teachers use inquiry-based learning in school,[32] evaluation showed that that this is not yet as common a practice as expected, despite the fact that this method is advocated by the Educational Department of the Dutch Government, and research skills are part of the official, national school curriculum. Some teachers in the evaluation panel reported back that the current programme is asking too much of their students, who are "not independent enough", "not used to this way of working" and with "questions too open". These teachers expressed a desire for a more directive approach, one that adheres more rigidly to step-to-step instruction, for both pupil levels (*havo* and *vwo*), particularly for those at the higher general secondary education (*havo*) level.

A need to differentiate between pupils of both levels is the second important insight. *Einstein was here* addresses both groups uniformly, a decision that was made as a result of the growing emphasis on open inquiry, which diminished the importance of their exact theory knowledge level. This decision is now being reconsidered: the initial idea of differentiated programmes for both levels is back on the table for development. Currently a slightly more directive instruction involving the replicas in the Lorentz Lab is being developed; still stimulating an inquisitive mind and tailored specifically to the level of higher general secondary education pupils.

As an activating teaching method *Einstein was here* contributes to the museum's mission to enable members of the public to make discoveries themselves, and – with the *Lorentz Formula* ending in 2022 – a comparable experience with working replicas is envisioned for general audiences in 2023. This programme is still in the early stages of development. Some of the school programme replicas, like the spectacular De La Rive tube, may be used again in this new programme. In addition, new replicas with a broader scope than electricity will certainly be added. Though not selected yet, requirements for these extra replicas are clear and challenging: easy-to-use instruments, capable of wowing the visitors in a swift-moving experience like the current *Lorentz Formula* and leaving the message that engagement with exact sciences is within the capabilities of everyone.

---

32   Many of them teaching at 'Technasia'; pre-university education with an extra emphasis on exact sciences and technique.

8   Conclusion

Developing the Lorentz Lab took over three years and has been an amazing undertaking as an experiment in itself for Teylers Museum, by being the first project to literally bring to life part of the history of the institute for visitors. The creation of a dedicated space has allowed the monumental rooms to remain unmodified, yet in active meaningful use, while also connecting audiences actively with the all-important underlying context of this heritage. The working replicas in the Lorentz Lab dramatisation specifically enhance the visitors' understanding of the original instruments that are on display, generating insight in how these instruments work, as well as why they are part of Teylers' history.

The Lorentz Lab, with educational and theatrical activities and a balanced act of evocation, action, thrill, pivots around the concept of a time capsule. This playing with visitors' sense of time has proven to be a very successful format for Teylers Museum. It has put a natural emphasis on the long timeline of the museum: travelling back in time makes visitors aware of this chronology and it enables them to become partners in this history rather than spectators. The working replicas in the historical setting strongly add to this experience. Presented as tools ready for scientific action, they allow visitors to participate directly in science and the big questions of the past. And when operated, the replicas reveal the original identity of the scientific museum collection in an accessible and attractive way. The review of the school programme didactics set aside, the Lorentz Lab has been deemed a successful and inspirational museum experiment by many. Developments along this line will continue in Teylers Museum, in the Lorentz Lab and other dedicated spaces the museum is preparing for the public.

CHAPTER 14

The Fall of Bodies according to Galileo

A Free Adaptation from the Geneva Museum of the History of Science

Stéphane Fischer | ORCID: 0000-0002-4928-9728

1   Introduction

For several years now,[1] the Geneva Museum of the History of Science has been developing educational modules in the history of science on themes related to its rich collection of historical scientific instruments. Intended for schoolchildren and families, these modules include demonstrations using historical instruments or modern replicas, as well as reconstructions of scientific experiments from the past. They cover a wide range of fields: acoustics, astronomy, electricity, electromagnetism, pneumatics, optics, and so on (Fig. 14.1).

Through this approach, the Museum offers its visitors a historical perspective that highlights the evolution of ideas and concepts rather than the more usual application of formulae. The intention is to raise public awareness of science through its history, particularly the history of physics which is the *raison d'être* of the Museum. One way to achieve that is to offer our audience spectacular experiments that show that the history of science is an exciting and relevant discipline.

The history of science and ideas is not often taught in Geneva or indeed in Switzerland as a whole. One exception is an introductory course to the experimental scientific approach given during the first year of *collège* (high school) in Geneva, which sometimes addresses certain aspects of the history of science. The Museum's activities thus complement school teaching of physics or astronomy. In this role, the Museum can rely on its collection of instruments, many of which are privileged witnesses to a discovery or conceptual breakthrough in the history of their discipline.

Workshops for young people generally last from 60 to 90 minutes. No preparation is required of participants. The challenge, therefore, is to capture their

---

1   This Chapter is based on the presentation: Stéphane Fischer, "The Fall of Bodies According to Galileo", *XXXIX Symposium of the Scientific Instrument Commission*, London, UK, 14–18 September, 2020.

© STÉPHANE FISCHER, 2022 | DOI:10.1163/9789004499676_016

FIGURE 14.1 Replica of the hemispheres of Magdeburg presented outside of the Museum in a workshop dedicated to the history of vacuum and atmospheric pressure, Geneva, 2010
© MUSÉE D'HISTOIRE DES SCIENCES DE GENÈVE

attention by making the visits as memorable and as lively as possible. Operation of the historical instruments is carefully prepared in advance and the accompanying explanation is well grounded scientifically. The demonstrations are more like 'shows' than standard physics courses, in which some of the historical experiments are presented in a more spectacular way than they were originally. The Museum does not reconstruct old experiments for research purposes, for example to verify whether or not they were feasible in the technical context of the time. Many historians of science are already engaged in this exercise.

In a more global perspective, historical presentations are an integral part of the mediation policy which has been undertaken by the Museum for several years including guided tours, lectures and plays that present science from a wide variety of perspectives. It is in line with the Museum's citizen approach which aims to offer visitors the keys to a better understanding of current scientific issues (global warming, the energy crisis, genetics, and so on).

In the context of a recent temporary exhibition entitled *Roulez les mécaniques, la loi du moindre effort* (Rolling Mechanics, the Law of Least Effort) devoted to the foundations of mechanics, the Museum has set up a new visit-workshop module dealing with a particular aspect of mechanics: the movement of bodies and in particular the fall of bodies. The workshop is freely

FIGURE 14.2　First page of a re-edition of Galileo's *Discourses* published in Bologna in 1655 and kept in the Library of the Museum of the History of Science in Geneva
© MUSÉE D'HISTOIRE DES SCIENCES DE GENÈVE

inspired by several experimental devices, notably the inclined plane and various pendulums, described or evoked by Galileo Galilei (1564–1642) in his famous *Discourses Concerning Two New Sciences*, published in 1638 at the end of his career (Fig. 14.2).[2]

2　Galileo Galilei, *Discorsi e dimostrazioni matematiche intorno a due nuove scienze attinenti alla mecanica et i movimenti locali,* Appresso gli Elseviri, Leiden, 1638; French translation:

2   Galileo, the Inclined Plane and Pendulums

Galileo published his last work during his 'house arrest' in Arcetri, near Florence. Written in the form of a dialogue between three characters (a scholar in favour of Aristotle's doctrine; a follower of experiments named Salviati, who is also Galileo's 'spokesman'; and an enquiring man seeking to understand), the book is considered one of the first modern treatises on mechanics. The Tuscan mathematician describes in detail two new sciences, the resistance of materials and local movement, which he developed at the beginning of his career in Pisa (1589–1592) and Padua (1592–1610). It is in this book that Galileo formulated his mathematical description of uniformly accelerated motion that is the basis of the fall of bodies.The dialogues are based on certain notions already developed by Galileo in his *Dialogue Concerning the Two Chief World Systems* published in 1632,[3] in particular the analysis of accelerated motion.

On the third day of the *Two New Sciences*, devoted to naturally accelerated motion, Salviati describes the experimental device used by Galileo to obtain proof that bodies in free fall do indeed fall in uniformly accelerated motion. "In a ruler, or more precisely a wooden rafter about 12 cubits long, half a cubit wide and 3 fingers thick, we cut a small channel just over a finger's width and perfectly straight; after lining it with a sheet of high gloss parchment to make it as slippery as possible, we let a very hard bronze sphere roll through it, perfectly rounded and polished".[4] Rather than dropping a body vertically, whose fall is far too fast to be measured, Galileo used a 6-metre long wooden inclined plane with a channel along which he rolled a bronze ball. The speed of the fall is slowed by the inclination which facilitates observations and, above all, time measurements. In Galileo's time, chronometers and watches did not exist. Measurements of short time intervals were made by counting the beats of the experimenter's pulse or by means of a pierced bucket the outflow from which, after a determined period of time, was weighed. Galileo's experimental approach consisted first of all of measuring the total duration of the fall along the inclined plane, then comparing it to the time needed to cover "half, or two thirds, or three quarters, or any other fraction; in these experiments repeated a good hundred times, we always found that the spaces covered were

---

G. Galilei, *Discours concernant deux sciences nouvelles* (trans. by Maurice Clavelin), Presses universitaires de France, Paris, 1995.

3   G. Galilei, *Dialogo sopra i due massimi sistemi del mondo, tolemaico e copernicano*, per G.B. Landini, Firenze, 1632; French translation: G. Galilei, *Dialogue sur les deux grands systèmes du monde* (trans. by René Fréreux), Editions du Seuil, Paris, 1992.

4   Galilei, *Discours...*, *cit.* (n. 2), p. 144, author's translation.

THE FALL OF BODIES ACCORDING TO GALILEO 267

between them like the squares of time, regardless of the inclination of the plane," observed Salviati.[5]

For a long time, science historians thought that Galileo had not conducted experiments on falling bodies and that his law of falling bodies was just an exercise in thought. However, in 1961, Thomas Settle, an American science historian, reconstructed the inclined plane described by Galileo and reproduced Galileo's experiments. He arrived at the same observations.[6] A few years later, the Canadian historian Stillman Drake, who immersed himself in the study of Galileo's manuscripts, suggested that Galileo used another method of measuring time that was much more accurate than measuring pulse beats or the weight of flowing water. Coming from a family of musicians and himself a lute player, Galileo would have used his sense of rhythm to measure very short intervals of time.[7] While studying an unpublished Galileo manuscript held at the Biblioteca Nazionale Centrale (Central National Library) in Florence, Drake discovered a series of numbers that would correspond to the consecutive distances travelled by the bronze sphere on the plane from its departure during eight regular time intervals. To define these time intervals, Galileo is believed to have placed lute gut strings through the drop channel so that they would vibrate and emit a sound as the ball passed through. By repeating the fall of the balls along the inclined plane, he would have finally managed to obtain regular time intervals between each pair of strings. He would then have noticed that the distances between each pair of strings (measured with a ruler) were unequal and increasing. Then, dividing the successive distances between two strings by the first distance, he would have discovered that the quotients corresponded to the sequence of odd numbers: 1, 3, 5, 7, 9, 11, and so on. This was a revolutionary discovery. For the first time, a physical phenomenon, in this case the fall of bodies, could be described by a mathematical law.[8] More recently, a team of researchers at the University of Oldenburg has also succeeded in reproducing Galileo's experiments on the inclined plane using his two means of measuring time.[9]

---

5   *Ibid.*, author's translation.
6   Thomas B. Settle, "An Experiment in the History of Science", *Science*, new ser. 133, 3445 (1961), pp. 19–23.
7   Stillman Drake, "The Role of Music in Galileo's Experiments", *Scientific American* 232 (1975), pp. 98–105.
8   Elizabeth Cavicchi, "Watching Galileo's learning", in L. Kime and J. Clark (eds.), *Exploration in College Algebra*, John Wiley & Sons, New York, 1997, pp. 581–595.
9   Falk Riess, Peter Heering and Dennis Nawrath, "Reconstructing Galileo's Inclined Plane Experiments for Teaching Purposes", Proceedings of the 8th International History and Philosophy of Science and Science Teaching (IHPST) Conference, Leeds, UK, 2006, available at: https://www.semanticscholar.org/paper/Reconstructing-Galileo%E2%80%99s

In his *Two New Sciences*, Galileo repeatedly describes experiments showing that the difference in velocity between two bodies of different mass decreases with the density of the medium being traversed. The denser the medium, the greater the difference of speed. In a low density medium such as air, the difference becomes almost imperceptible. Galileo imagined the daring hypothesis that if "the resistance of the medium was totally eliminated, all the bodies would descend at the same speed".[10] In Galileo's time there were no vacuum pumps. It was therefore impossible to experimentally verify his assumption which ran counter to the classical conception of the Universe in force at the time according to which heavy bodies fall faster than light ones.

However, Galileo tried to prove his assertion. One of the experiments consisted of dropping bodies of different mass and then observing which touched the ground first. But the duration of the fall is often very short and does not allow very detailed observations, so Galileo resorted to another experimental procedure: rather than observing a single fall, he repeated falls using a bifilar pendulum.[11] This device consists of two balls: one of lead and one of cork, each suspended at the end of two 2.5 m long wires so that their plane of oscillation remains constant. The two balls are released simultaneously. After a hundred beats, Galileo found that the lead ball did not beat faster than the cork ball and that the two balls maintained exactly the same movement. The cork ball loses a little in amplitude because of friction but still oscillates at the same period as the lead ball.[12]

While studying the pendulums, Galileo discovered that the period of a pendulum does not depend on the suspended mass but only on the length of the wires. He formulated his find in the following way: "As for the times of oscillation of mobiles suspended at the end of wires of different lengths, they have between them the same proportion as the square roots of the lengths of these wires, which means that the lengths are between them like the squares of time".[13] Unfortunately, Galileo does not give any precise description of the pendulums used for his observations. Several science historians have set out to 'reinvent' pendulums similar to those Galileo could have used to highlight

---

-Inclined-Plane-Experiments-Riess-Heering/2cc437dd5002913f7e84142e188ecfc39 04a3048 (accessed 9 Mar. 2021).

10 Galilei, *Discours...*, *cit.* (n. 2), p. 61, author's translation.

11 Robert Signore, *Histoire du pendule*, Editions Vuibert, Paris, 2011, p. 6.

12 Galilei, *Discours...*, *cit.* (n. 2), p. 70.

13 *Ibid.*, pp. 78–79, author's translation.

THE FALL OF BODIES ACCORDING TO GALILEO

certain properties,[14] notably the dependence of the period on the length of the wire and its isochronism, *i.e.* that during weak oscillations, the period does not depend on the amplitude.

3   The Inclined Plane of the History of Science Museum

The Museum's inclined plane is based on Galileo's description of it in his *Two New Sciences* and on a reconstruction presented in 2005 in Geneva as part of an exhibition devoted to Galileo's work during his stay in Pisa.[15] It was made by the team of carpenters of the Natural History Museum of Geneva. For practical reasons of storage, the inclined plane is made up of two sections of 330 cm each, which are assembled with screws. It rests on two wooden legs, one of which is adjustable so as to change the inclination. The sloping top is equipped with a gutter along which 20 mm diameter metal ball bearings roll. A graduated tape is glued to one side of the plane to measure the distances travelled (Fig. 14.3).

Inspired by the inclined plane described by Jean-Antoine Nollet in his *Leçons de physique expérimentale*,[16] and by a nineteenth-century model kept in the Museo Galileo in Florence,[17] our inclined plane is equipped with movable bells that ring as the balls pass and can be moved along the fall channel. The bells replace the lute-string casings that Galileo installed on his inclined plane. They are essential elements of the experimental set-up (Fig. 14.4).

Curiously, we found very few detailed descriptions of inclined planes in eighteenth- and nineteenth-century physics treatises. In his *Leçons de physique expérimentale*, Nollet depicts an inclined plane in the form of an inclined rope along which slides a small weighted ball with a spike. Bells are arranged along a second rope aligned above the first. They ring when they are struck by the spike of the moving ball. Time is measured by a pendulum.

---

14   Paolo Palmieri, "Experimental History: Swinging Pendulums and Melting Shellac", *Endeavour* 33, 3 (2009), pp. 88–92.

15   See the exhibition catalogue: Roberto Vergara Caffarelli, *Galileo e Pisa*, Felice Editore, Ospedaletto, 2004, pp. 44–47, 62–65.

16   [Jean-Antoine] Nollet, *Leçons de physique expérimentale*, chez les fréres Guerin, Paris, 1749–1755, 4 vols.; v. 2 (1753), pp. 161–164.

17   Inclined plane, early nineteenth century: Museo Galileo: Institute and Museum of the History of Science, Florence, inv. no. 1041.

FIGURE 14.3   The inclined plane of the Museum, 660 cm long. A graduated tape is glued to one side of the plane to measure the distances travelled by the rolling balls
© MUSÉE D'HISTOIRE DES SCIENCES DE GENÈVE

Other works, such as René Just Haüy's *Traité élémentaire de physique*,[18] describe the law of the fall of bodies according to Galileo's geometry. In a right-angled triangle horizontal lines indicate increases in speed during equal time intervals (vertical line separated by equal segments); the surface of the triangle is the sum of the spaces travelled by the mobile during a given time. Finally, some nineteenth-century physics textbooks (as those by Alphonse Ganot and Amédée Guillemin) give only a very succinct explanation of the inclined plane with a simple diagram representing the different vectors of forces acting on the falling ball.[19] On the other hand, considering the inclined plane experiment as "not giving results of great precision",[20] these books describe in much more detail other devices that appeared in physics cabinets as early as the eighteenth century to demonstrate the law of falling bodies, such as Newton's tube, Atwood's machine or Morin's apparatus.

---

18   René Just Haüy, *Traité élémentaire de physique*, Delance et Lesueur, Paris, 1803, 2 vols.: v. 1, pl. 1, fig. 1.
19   See: Alphonse Ganot, *Traité de physique*, Hachette, Paris, 1884, pp. 76–77; Amédée Guillemin, *Les phénomènes de la physique*, Hachette, Paris, 1869, pp. 23–24.
20   *Ibid.*, p. 24.

THE FALL OF BODIES ACCORDING TO GALILEO 271

FIGURE 14.4   Detail of a bell installed on the inclined plane to tinkle at the passage of the ball in the channel
© MUSÉE D'HISTOIRE DES SCIENCES DE GENÈVE

Newton's tube (named after the famous English natural philosopher) is a glass tube that can be connected to a vacuum pump to evacuate its air. Two bodies of different mass, for example a feather and a glass ball, are placed in it. The two objects fall at the same speed when the tube is emptied of its air and turned upside down. When air re-enters the tube, the feather slows and falls less quickly than the ball. Atwood's machine (named after its inventor) consists of a large wooden column 2.5 m high topped by a large pulley with a

FIGURE 14.5   Demonstration of the accelerated movement of the balls rolling along the inclined plane. The bells placed at equal distances from each other emit a tinkling sound closer and closer together
© MUSÉE D'HISTOIRE DES SCIENCES DE GENÈVE

groove, equipped with very sensitive bearings. A fine rope supporting an identical weight at each end passes over the pulley. When a slight force is exerted on one of the weights, it begins to descend vertically, slowed down in its fall by the inertia of the other weight. Morin's apparatus is composed of a large paper-covered rotating cylinder more than 2 m high on which the falling body, equipped with a pencil lead, traces its trajectory. Both Morin's device and Atwood's machine provide the means to slow down the fall without altering the nature of the movement.[21]

At our museum, during one of the introductory experiments with the inclined plane, the demonstrator asks the audience to observe the behaviour of the falling ball. The ball has zero initial speed. What happens during its descent? Does the speed increase or does it remain constant? And how can this be verified? Reminding the participants that Galileo did not have a stopwatch accurate enough to measure short intervals of time, and so had to resort to a system of bells, they are asked to describe how the bells were used in our experiment. After a short discussion, the answer is relatively quick: the bells are placed at regular intervals of distance along the plane. By rolling the ball, we can clearly hear the rings becoming closer and closer together, indicating that the ball is accelerating during its fall (Fig. 14.5).

---

21   Ganot, *Op. cit.* (n. 19), p. 80.

THE FALL OF BODIES ACCORDING TO GALILEO

The demonstration continues by repeating in a very crude way the experiment described by Galileo in his *Two New Sciences*. With a stopwatch and a bell placed at the lower end of the inclined plane, a volunteer from the audience is invited to measure the total duration of the fall. She or he starts the stopwatch on releasing the ball and stops it at the sound of the bell. The demonstrator then asks him or her to move the bell to the place that would correspond to the time of half the fall. In the majority of cases, the bell is placed at half distance. Curiously, the previous experiment with bells placed at equal intervals showing accelerated movement of the ball from a zero initial speed, does not seem to be retained by the participants. In people's minds, the ball seems to move at a constant speed. A few quick measurements of the travel time by the volunteer with the help of the demonstrator at half, a quarter and a ninth of the distance, shows that the ball is not moving in uniform rectilinear motion but is accelerating. It is at one-quarter of the length that the travel time is halved, and at one-ninth of the length that the travel time is one-third as measured by the stopwatch. Comparing the times registered on the one hand and the distances travelled on the other, we arrive at Galileo's observation that "the spaces travelled are like the squares of times".[22]

The slope of the Museum's inclined plane can be adjusted to different angles. If we had more time in the workshop, we could repeat the experiment by varying the slope and show that the movement of the ball always accelerates naturally, independently of the slope. If we had an inclined plane with a horizontal table placed at its lower end, we could reproduce a uniform rectilinear movement by rolling the ball onto it.

Finally, this museum demonstration pays tribute to Galileo's musical ear in his work. First, two bells are placed by the demonstrator at predetermined distances so that they each ring at equal time intervals. A spectator is then asked to place the third bell so that it rings at the same rhythm as the two previous ones. The experiment can of course be extended with a fourth bell. Once the bells are placed on the plane, an electronic metronome is used to check that the rhythm of the ringing is constant. The distances between each bell are then measured using a graduated ruler, and especially the relationship between the successive distances between two bells and the first distance, according to the method of calculation adopted by Galileo. So, we discover that the quotients obtained evolve as a sequence of odd numbers: 1, 3, 5, 7, and so on (Fig. 14.6).

---

22   Galilei, *Discours...*, *cit.* (n. 2), p. 144; author's translation.

FIGURE 14.6   The volunteer experimenter concentrates on releasing the ball at the beat of the electronic metronome
© MUSÉE D'HISTOIRE DES SCIENCES DE GENÈVE

4 Other Experimental Devices of the Educational Module

While the inclined plane is the centrepiece of this workshop, it is accompanied by other demonstrations designed to better understand the phenomenon of falling bodies. By way of introduction, the mediator asks the audience which of two falling balls, a steel *petanque* ball or a polystyrene ball of the same size, would be the first to reach the ground. The workshop participants share their answers. Half the audience is in favour of the steel ball because it is heavier. The other half thinks that weight has no influence and that both balls would touch the ground simultaneously. Ideally, the experiment should be repeated from a greater height, for example from the roof of the Museum, as is sometimes the case during our outdoor demonstrations. In this case, the steel ball clearly hits the ground first. Failing that, we drop the balls from a stepladder. We see that the steel ball touches the ground first. This appears to confirm the Aristotelian view of the world whereby heavy bodies fall faster than light ones. We then force the issue by repeating the experiment with the steel ball and a feather. Here, clearly, the feather is still floating in the air when the ball has already touched the ground. If one now asks why the feather falls so slowly, the

THE FALL OF BODIES ACCORDING TO GALILEO   275

audience answers that it is because it is lighter, but also because of the resistance of the air. This introduces Galileo's bold claim that without air resistance, the feather (and the polystyrene ball) would fall as fast as the steel ball.

The audience is then presented with a bifilar pendulum inspired by the one described by Galileo in *Two New Sciences*. It is explained that he had designed this device to reproduce a series of falls, consisting of the oscillations of the pendulum – much easier to observe than the simple fall of a body thrown from any height. Two balls of the same size, a heavier one in steel and the other lighter in glass, are hung from the ends of two 80 cm long suspension wires. The pendulum is released by a spectator. After a dozen or so back and forth movements (not a hundred, as documented by Galileo), the two balls still oscillate at the same rhythm although the glass ball has lost a lot of amplitude. The phenomenon can also be verified by the audience with another device (inspired by a reconstitution of Galilean pendulums made by the late Thomas Settle, while collaborating with colleagues at the Department of Physics of the University of Padua).[23] Four pendulums of identical length but made of spheres of different materials (paper, wood, glass and steel) are hung at the end of a frame. By releasing them simultaneously, the four pendulums oscillate at the same rhythm for a certain period of time.

The following question is then put to the public. If the period of a pendulum does not depend on the suspended mass, what can modify it? It is hypothesised that it could be related to the length of the wire to which the weight is attached. The demonstration is repeated with the other part of the device. This time four steel balls of identical weight are suspended from wires of 100 cm, 50 cm, 25 cm and 11.11 cm. The 100 cm long pendulum is used as a reference. We then ask the audience which pendulum would beat twice as fast, the 50 cm, 25 cm or 11.1 cm. The most frequent answer indicates the 50 cm pendulum. Having simultaneously launched the 100 cm and 50 cm pendulums, the demonstrator asks a spectator to count the number of beats of the 50 cm pendulum while the reference pendulum makes 10 beats (counted by the demonstrator). The experiment is repeated with the reference pendulum and the 25 cm pendulum. This time, the observer counts 20 beats, that is to say, double that of the reference pendulum. The last question is in the form of a game of logic. The audience is asked how fast a pendulum nine times shorter than the reference pendulum will beat, knowing that a pendulum four times shorter beats twice as fast. The answer is verified experimentally by simultaneously launching the reference pendulum and the pendulum of 11.11 cm. The audience is reminded

---

23   Enrico Bellone, *Galilée, le découvreur du monde*, Editions Belin, Pour la science, Paris, 2003, p. 52.

FIGURE 14.7   Demonstration of the law of the pendulum establishing that the oscillation period depends only on the length of the wire
© MUSÉE D'HISTOIRE DES SCIENCES DE GENÈVE

that the relationships between the lengths of the wires and the period of the pendulum, are similar to those observed between the running time of the ball and the lengths travelled on the inclined plane (Fig. 14.7).

Our workshop ends with two demonstrations designed to verify Galileo's intuition that in a vacuum all bodies fall at the same speed. The first is performed by the demonstrator using a modern replica of a Newton tube. This demonstration extends the story of the fall of bodies in Newton's work and reminds the public that it was, among other things, on the basis of Galileo's observations that the English scholar succeeded in expressing mathematically the famous law of universal gravitation (Fig. 14.8).

Finally, we show visitors a short film dating from 1971, of what is probably the most expensive experimental demonstration ever made of Galileo's law of the fall of bodies. One of the astronauts on the Apollo 15 mission to the Moon (where there is almost no atmosphere) holds in his hands a hammer and a feather that he releases simultaneously. Both objects touch the ground at the same time.[24] The screening of this short film is a light-hearted way of concluding

---

24   *Apollo 15 Hammer-Feather drop*, 1971, nssdc.gsfc.nasa.gov (accessed 4 Mar. 2020).

FIGURE 14.8   Demonstration of falling bodies in vacuum with a modern replica of a Newton tube
© MUSÉE D'HISTOIRE DES SCIENCES DE GENÈVE

the workshop. It also serves as a reminder that the verification of the law of falling bodies is still relevant today.

5 Review and Conclusion

Our educational module on falling bodies lasts about 60 minutes. It is both very long and very short. Very long when you consider that the subject is not the most attractive. How can we interest an audience that is most often schoolchildren (primary and secondary) in a theme that is not very promising compared to the subjects of our other workshops, such as electricity or magnetism, which feature electrostatic machines or induction coils that produce sparks and electrical discharges that are much more spectacular? Like all the activities offered to the public, the module described here focuses on practical demonstrations. It is led by a mediator who explains the objectives and places them in their historical context. The demonstrations require both great manual dexterity and a mastery of intellectual content. The audience does not remain passive. They are invited to take part in the experiments in various ways: releasing the balls and measuring the time they take to travel along the inclined plane with a stopwatch; moving the bells on the inclined plane so that the rhythm of their ringing matches that of a metronome; setting the various pendulums in operation; counting the number of oscillations, etcetera. These activities usually generate a strong active response from the public. They may also realise how difficult it is to make apparently very simple reliable measurements such as measuring the fall of a ball or counting the number of beats of a pendulum.

But the module is too short if one wants to go deeper into certain aspects of the history of the law of falling bodies. For example, the Aristotelian conception of the Universe is often unjustly caricatured, but Galileo himself was still imbued with some of Aristotle's notions such as the impetus, a kind of momentum communicated to bodies during an unnatural movement. Similarly, the mathematical approach based on Euclidean geometry also deserves further explanation. One could also extend the discourse to Einstein and his principle of equivalence between gravitation and acceleration during the fall of bodies into a vacuum at the origin of the theory of general relativity. These more conceptual themes could be integrated into workshops aimed at a more educated audience at the college and university level, where those physics subjects are addressed in the curriculum. Such workshops, which would last for several days, could be conducted in partnership with physics teachers. The practical part could take place in the Museum and the theory in the colleges and universities. For the time being, this kind of educational project is still at a draft stage.

For the team at the Museum of the History of Science, the inclined plane is a formidable didactic instrument. On the one hand, it allows us to apprehend a physical phenomenon that is rarely approached in a practical way in schools and colleges and, on the other, it is a symbol of a modern scientific approach. It does not matter whether the inclined plane served as an experimental validation of Galileo's concepts or whether, on the contrary, it inspired him in his reflections and conclusions. During our workshop, the inclined plane becomes almost a totem, an icon of what modern science is: hypothesis, measurement, experimental validation, and so on.

Our approach aims to introduce the public to both the scientific approach and to Galileo's work. To take up this challenge, we must necessarily go through a few historical shortcuts and schematise Galileo's thought and reflection. But it is impossible to leave the inclined plane as a static public display. This instrument only reveals its full experimental power when placed in its historical context. This implies a good knowledge of the subject on the part of the moderators and transmission of selected basic information to visitors, as well as a level of adaptation to the public.

More generally, after the workshop devoted to the fall of the bodies, the inclined plane becomes in the eyes of the visitors and the teachers a flagship instrument of the Museum next to the machine for reproducing the polar aurora or the great eighteenth-century orrery. The workshop on the inclined plane fulfils one of the three official missions of a museum, which is the promotion of its heritage (material or immaterial) to the public. The other two missions are conservation and scientific study. At the Museum of the History of Science, this promotional mission can take different forms: lectures, guided tours, interactive workshops, theatre pieces, and so on.

All these activities play an essential role in the visibility of the Museum within society, widening its renown and attracting new audiences. This can only be achieved by proposing activities centred around the Museum's own resources, which are its collections. Our museum is a museum of the history of science and not of science. The nuance is important. The collections are our wealth and our trademark. All the activities that we propose have a close link with the instruments of the collections which are formidable witnesses of the history of concepts, sciences and techniques.

Index

1st Upper High School (Lyceum) Vrilissia, Athens, Greece 181
26th General Lyceum of Athens, Greece 108
39th Primary School of Athens, Greece 119
3D 53, 116, 130, 137–138, 146
  film 146
  printer 130, 137–138
  scan 138
  view 79
4th Lyceum of Palaio Faliro, Athens, Greece 181–182, 208

a priori 224
a-didactic 214
abstract reasoning 17
Academy of Sciences, France 94, 123
Accademia del Cimento, Florence, Italy 187
accelerated motion 266, 273
Accompagnement en Science et technologie à l'école Primaire (ASTEP), Rennes, Frances 96–97
acetylcholine 73
acid 38
acoustic 2, 44, 77, 85, 115–116, 142, 152
acoustic siren 2, 44, 115–116
acoustics 7, 142, 263
actinometer 86, 91
action research 62
actor 103, 181, 200, 247–248
adenoid curette 55
Aditomo, Anindito 61
aesthetic 42, 57, 167, 179
affordance 17–21, 27
affording 17–18
Africa 12
African-American 167
age (of person) 1, 12, 75, 82, 95, 108, 128, 181, 234
agency 11, 33, 161, 180
air pump x, 66, 69, 192, 194, 202, 230
air-pump receiver 192
airplane wing 70
algebraic curves 10, 218, 223
algebraic formulas 217–218
alidade 24

American, *See* United States of America
amis du musée historique du Lycée Hoche, Les Versailles, France 128
ammeter 126
Ampère table 89
Ampère table modified by Bertin 89
Ampère's law 255, 257
Ampère, André-Marie 89, 125, 255, 257
anachronism 43, 253
anaesthesia mask x, 56
analogues 2, 6, 163, 171, 255
anamorphised object 143, 146
anatomy x, 55–56, 75
ancient vii, 31, 36, 126, 217
Angelini, Pierre-Antoine 91, 104
angular speed 144–145
anorthoscope 2, 142–146
anthropology 28, 54–55
antiquarian vii
aperiodic balance 99
aperture 53
Apollo 15 276
Arab 12
Arabic language 33
Arago, Dominique François Jean 89
arc singing, speaking 44, 47
Arcetri 266
archaeology 10, 50–52, 61, 139, 207
Archief en Documentatiecentrum Nederlandse Gedragswetenschappen Gedragswetenschappen – ADNG 80–81
Archimedes 95, 127, 130–131, 150, 217
Archimedes' principle 127, 150
Archimedes' screw 95, 130–131
architecture 19, 54, 174–176, 178
  institutional 19
Arduino 73–74
Aristotle 17, 19, 28, 266, 274, 278
armillary sphere 154
art educator 21
artificial lightning 195
artisan 88, 163, 239, 240
asbestos viii
ASEISTE, *See* Association de Sauvegarde et d'Étude des Instruments Scientifique et Techniques de l'Ensignement

Asia 12
assessment 14, 242
assimilation and accommodation 214
Association de Sauvegarde et d'Étude des
    Instruments Scientifique et Techniques
    de l'Ensignement (ASEISTE) 5–6,
    122–138
Association des Physiciens de Lyon (APL)
    133
Association for Preserving and Studying the
    Scientific and Technical Instruments of
    Education (ASEISTE), *See* Association de
    Sauvegarde et d'Étude des Instruments
    Scientifique et Techniques de
    l'Ensignement (ASEISTE)
Association Science en Seine et Patrimoine
    (ASSP), Rouen, France 211
astrolabe 24, 30, 158, 161–162, 171–172,
    174–179
astronomer 24, 167
astronomy 12, 31, 140–141, 154, 156, 160, 168,
    178, 186, 203, 263
Athens 6, 105–121, 181–208
Athens Science Festival 2016 181, 183,
    202–203, 207
Athens Teaching School 105–107
Athens University History Museum
    198–200, 207–208
atom 14, 24, 142
Atwood's machine 270–272
Augustus the Strong of Saxony 37
authentic 5–8, 111, 140–141, 163, 175, 179,
    252–254
automated instruments 16
autonomous 109, 115, 141, 209, 214
autonomy 9–10, 214, 222, 224–225
average 126
axiom 175, 217

bachelor (undergraduate) 139, 212–214
bachelor project 9, 140–143, 145, 149,
    153–154, 156–157
Baconian tradition 188, 195
Baekeland, Leo 139, 152–153
Bakelite 126, 152
balance (weighing) 167–168
Bales, Jim 180
Balfousias, Panagiotis 190, 193
Balomenou, Katerina 182

Baltic-German x, 55–56
Banneker, Benjamin 168, 170
Bareha, Younès 88
Barlow's wheel 256, 258
barometer 187–188, 193–195
baroscope 127
battery 38, 44, 47, 79, 113, 197, 254
battery of Leyden jars 47
bead lens 158, 163
Beaudouin, Denis 99
Beaulieu Scientific Campus, Rennes,
    France 85, 90
beautiful rescues 102
Belgian, *See* Belgium
Belgium vii, 82, 204
bell (timing) 4, 269, 271–273, 278
Bell and Siemens 105
Benjamin, Walter 175
Bernard, Dominique 7, 84–104
Bernoulli family 70
Bernoulli, Daniel 70
Bernoulli, Jakob 70
Bernoulli, Johann 65–67, 70, 125
Berti, Gasparo 187–188, 193–194, 208
Bertin 89
Bertozzi, Eugenio 47
bibliography 85, 91, 190, 215, 219
Biblioteca Nazionale Centrale, Florence
    267
bifilar pendulum 268, 275
Bingen, Hildegard von 82
binocular vision 146
biology 32, 75, 82, 102
bioscope 142, 146–149
Biot, Jean-Baptiste 89
black box vii, 48, 103, 259
Bliuc, Ana-Maria 61
bloodletting 30
Bloom, Benjamin 32
blower with bellows 44
blowpipe 164, 179
bodily 2, 4, 11, 19–20, 29–31
    discomfort 20
    experience 3, 11
    practice 30
body-object relationship 19
boiler 38, 126
Boston, Massachusetts 165–166, 172–173,
    179–180, 241

INDEX 283

Bourbouze galvanometer   132–133, 135
box-plot   234–236
Boyle, Alison   13
Boyle, Robert   182, 187, 189, 192, 202
brachistochrone   70–72
Braun's tube   38, 196–197
Brenni, Paolo   x, 7–8, 15, 34–49, 97, 123
briki   20
Britain, *See* United Kingdom
Brittany, France   84–85, 87, 94, 104, 126
Brohinsky, Jais   169–171, 180
budding guide programme   130
build instrument   60, 77, 91, 184, 212, 213, 219,
     220, 222, 223, 226, 239–241, 256
Bunsen burner   44, 164, 166
Bunsen, Robert   188

cabinet de physique des enfants de France,
     Le   128
Cabinet of Curiosity, Brest, France   10,
     209–225
cabinet of physics, The (Joseph Plateau),
     Ghent, Belgium   8, 66, 143, 150
Cahier de Ma Vie en Prison, See Helsen, Josef
calculating instruments   91, 224
calculator   24, 91, 97, 172, 174, 209, 212, 219,
     224
Caloggero, Tony   180
calorimeter   94
calorimetric experiment   42
Cambridge, Massachusetts   176
camera   42, 45, 60, 237–239
     photo   42
     video   42
camera obscura   6, 158, 237–241
Canada   61, 267
Candlin, Fiona   59
Caplan, James   vii
Carchon, Roland   8, 139–157
carpentry   128, 241, 269
Cartesian method   218
Carthusian monks   232
cartography   153
cartoon film   146
case study   1, 5, 8, 11–13, 19, 32, 51, 60–61, 184,
     226
case-based   61
catalogue   30, 39, 57, 68, 77, 83, 110–111,
     115–116, 122–123, 163, 246, 249, 269

cathode ray   186, 188, 196
caution   vii, 28
Cavicchi, Elizabeth   vii, 1–13, 64, 158–180,
     208, 241
celestial sphere   175
cellphone   161, 171–172
Centre for Public Engagement, Groningen,
     Netherlands   83
Centre François Viète, Brest, France   88
cephalograph   x, 55–56
chalice   36
Chambet, Audrey   89
Chaplin, Charles   124, 203
Charles River Bridge   167, 179
Charles X, King of France   128
Chatterjee, Helen J.   51, 58–60, 62
chemistry   viii, 7, 38, 43, 54–55, 57, 73, 75,
     85, 87–90, 93, 100–102, 105, 123, 128, 131,
     152, 186, 203, 209
Chiavacci, Antonio   35, 45
     children of France's physics cabinet, The,
     See cabinet de physique des enfants de
     France, Le
chimney   70, 228–230
Chinese   33, 167–168
chromium dichromate   38
chronometer   266
cinematography   42
cissoid   217
citrus reamer   19
civil defence siren   115
civil engineering   10, 168–169, 179
Clarke dynamo   254, 256, 258
class (social)   1
class visit   10, 162–163, 167, 172, 203–204
classics   57, 206
classification   14, 90, 186, 217
clock   36, 77, 158, 218
cloud, data   97
CNAM, *See* Conservatoire National des Arts
     et Métiers (CNAM) – Museum des Arts et
     Métiers
Cobi, Alban   180
Coca-Cola   28
coffee   20
collaboration   viii, 3, 5–7, 9–10, 12, 34–35, 47,
     50–51, 59, 85, 90, 104, 116, 118–119, 122,
     129–130, 138, 158, 160, 170, 181–182, 186,
     190, 201–203, 242, 248, 257, 275

284 INDEX

collapsible compass 217
collection vii, 2–4, 6–8, 10–11, 30, 33–34,
  38, 42, 48–49, 50–52, 54–57, 60–62,
  65–67, 69–73, 75–77, 79, 81–84, 86,
  88–91, 93–95, 97, 99, 101–102, 105–109,
  111, 113–115, 117, 120, 123, 126–129, 131,
  133, 139–140, 142–144, 150, 152–157, 161,
  163, 198–199, 207, 209, 211–212, 223, 237,
  240, 243–247, 250, 262–263, 279
  manager vii
  specialist 3, 71
  See Archief en Documentatiecentrum
    Nederlandse Gedragswetenschappen
    Gedragswetenschappen – ADNG
  See cabinet de physique des enfants de
    France, Le
  See Cabinet of Curiosity, Brest, France
  See cabinet of physics, The (Joseph
    Plateau), Ghent, Belgium
  See Collection for the History of Sciences,
    Ghent, Belgium
  See Harvard University, The Collection of
    Historical Scientific Instruments
  See University Libraries Special
    Collections, Groningen, Netherlands
  See Van Heurck collection, Antwerp,
    Belgium
Collection for the History of Sciences, Ghent,
  Belgium 139, 154, 156
collector vii, 48, 128, 162, 209–211
college vii, 21, 58, 72–73, 84, 94, 126–127, 153,
  159, 211, 267, 278
collège 263
Collège du Château, Morlaix, France
  126–127
commitment 13, 217
communal contribution 33
commutator 38
compass viii, 167, 171–174, 179, 217, 219, 223
  Galileo (geometrical) viii, 167, 171–174,
    179
  See collapsible
  See divider
  See drawing
  See hypothetical
  See magnetic
  See proportional
competence 13–14, 214
compressor 44

computer 16, 29, 37, 48, 72, 77, 79, 97, 145,
  221–222
  laboratory 53, 63
  science 85, 91, 140
  screen interface 17
computer-controlled 72
computer-generated image 79
computer-generated material 79
conceptual knowledge 3
conchoid
  of Nicomedes 217
conics 217, 219
  sections 219
connoisseurship 29
Connors, Alanna 180
ConScience 130
conservation 34, 37, 53–54, 85, 89, 94–95,
  101, 120, 279
Conservation and Digitisation Centre Kanut,
  Estonia 53
Conservatoire National des Arts et
  Métiers (CNAM) – Museum des Arts et
  Métiers 85
construction 1, 75–77, 80, 142–143, 146, 149,
  154, 173, 195, 198, 203, 215–220, 222–223,
  252
contact (electrical) 38, 255
contemporary heritage 88
Contemporary Scientific and Technical
  Heritage, See Patrimoine Scientifique et
  Technique Contemporain (PATSTEC)
context-aware functionality 24
contexts 1, 7, 11, 33, 47, 186
  culturally-sensitive 1
contextual science 181, 184–185, 118–208
cooperation 52, 59, 64, 73, 75, 83, 116, 253
Cordeiro-Antunes, Gabriela 180
Corfu school 106
Cotton balance 134, 136
Couch, Alva 180
Coulomb's Law 31
counterweight 21
Covid-19 3, 89, 181
craniometer x, 55
creativity 4, 119, 159, 198, 214
Creator 68
critical exploration in the classroom 159
critical thinking 32, 213, 259
Crookes' radiometer 203

INDEX 285

Crookes' tube   196
Crookes, William   31, 38, 188, 196–197, 203
crown of cups   233
cryostat   101
crystallochemistry   88
crystallography   88–90
cube's duplication   217
cultural   vii–viii, 1, 3, 12–13, 17, 28, 30–31, 51, 62, 75, 83, 90, 95, 104, 112, 125, 132–133, 189
   heritage   viii, 132–133
   norms   17
   value   30
culture   vii–viii, 1, 3, 11–15, 17, 28–31, 33, 51, 62, 75, 83, 87, 90, 95–97, 104, 112, 121, 125–126, 132–133, 162, 167, 183–185, 189, 202, 207–208, 241
   See material
Culture and Heritage team, Rennes, France   87
culture of doing   31
curator   vii, 3–5, 7, 32, 52, 54–55, 57–58, 60–64, 71, 95, 98–100, 162, 173–174
Curie devices   98–99
Curie scale   87
Curie, Jacques   98–99
Curie, Marie   7, 84, 98–100
Curie, Pierre   7, 84, 98–100
Curie-Blondlot wattmeter   98–99
Curie-Cheneveau magnetic balance   98–99
curiosity   2, 5, 10, 36, 95, 103, 161, 163, 166, 174, 177, 209, 212, 215, 224–225, 254
curriculum   10, 14, 30, 32, 51, 63, 77–78, 105, 119, 125, 128, 131–133, 140, 156, 159, 184–186, 190, 206, 223, 225, 248, 261, 278
Cuthbertson, John   37, 251–252
cycloid   123

d'Alembert, Jean-Baptiste le Rond   125
Dachau   204–205, 208
   concentration camp   205
   concentration camp memorial   204
   exhibition   205
Dale, Sir Henry Hallett   73
Dalloubeix, Christine   128
damage   38, 41–42, 68, 109, 115, 150, 226
dangerous   viii, 24, 38, 199, 201, 230, 254
data collection   114–115
database   52–56, 63–64, 87, 89

Dauphin, Louis of France   128
Davy lamp   111, 114–115
De Carvalho, Renata   166, 179–180
de Kuijper, Ewie   254
De La Rive tube   249, 255, 258, 261
de Serent, Marquis Armand-Louis   128
de-contextualisation   1, 4, 227
deformed image   144
demonstration   190, 196, 199
Department of Physics, University of Padua, Padua, Italy   275
Deprez-D'Arsonval galvanometer   132–133
Descartes, René   217–218
   Géométrie   217, 219
design   x, 1–2, 11–12, 15–17, 19, 25, 28–29, 31–33, 50–51, 53, 57–58, 61, 63–64, 83, 94, 111, 115, 120, 144, 173–174, 178, 184–185, 190, 195, 199, 212, 214, 218, 228, 250, 252, 257
   analysis   28–29, 31–33
   theory   16, 25
desktop computer   96–97
deterioration   viii, 94
Deutsches Museum, Munich, Germany   203
Devoy, Louise   13
Dewey, John   32–33
Deyrolle, Emile   89
diagonal rod   221
didactic   21, 31, 37, 45, 48, 210–211, 213–214, 223, 257, 260, 262, 278
diffraction and secondary electron microscopy   151
diffraction chambers   88
digital voltmeter   151
Dimacali, T.J.   180
directive instruction   261
director   vii, 77, 164, 245
directrix   219, 221, 223
disability   1, 16
discourses   13, 265
dissemination   96–97, 132–133, 189
divider compass   217
do-it-yourself (DIY)   209, 212, 215
doctoral thesis   5, 88, 105, 152
documentation   3, 27, 30, 90, 102, 104
domestic objects   52
Donderkerkje (thunder church)   68
Douglas, Deborah   162–163, 165, 167–169, 171, 180

Drake, Stillman   171, 187, 267
drama   6, 247–248, 262
Draper, Henry   151
Drummond lamp   123–124
dry box   88
Duboscq disc   148–149
Duboscq prism   87
Duboscq, Jules   x, 43, 87, 142, 146–149
Duckworth, Eleanor   159, 180
Ducretet, Eugène   86, 89, 108
Dutch   66, 75, 81, 249, 252–253, 261
Dutch Archive for the Behavioral Sciences,
   See Archief en Documentatiecentrum
   Nederlandse Gedragswetenschappen
   Gedragswetenschappen – ADNG
dynamo machine   44

e/m   143, 196, 198
earthquake   106–107
École Supérieure d'Ingénieurs de Rennes
   (ESIR), Rennes, France   91
ecological   16–17, 32–33
economic status   1
Eddington, Arthur   31
edge-on   24
Edgerton, Harold   160
Edison phonograph   105
editorial board   vii
education theory   16
educational   vii–viii, 3, 5–13, 33, 47, 51, 65,
   69, 72, 96–97, 100, 103, 108–109, 114–117,
   120, 124, 128, 139–140, 156, 158, 172, 184,
   190–191, 196, 198, 204–205, 208, 213–214,
   225–227, 237, 242, 247–248, 256, 258,
   260–263, 274, 277–278
   activity   vii–viii, 9, 12, 51, 120, 188, 247
   initiative   vii
   purpose   69, 72, 124, 139–141, 143, 145, 147,
      149, 151, 153, 155, 157
   vision   3
Educational Department of the Dutch
   Government   261
educational programme   109, 114–115, 117,
   248, 256, 260
   See Accompagnement en Science et
      technologie à l'école Primaire (ASTEP),
      Rouen, France
   See Athens Science Festival 2016
   See bachelor (undergraduate)

See Centre for Public Engagement,
   Groningen, Netherlands
See European Day of University
   Collections
See European Heritage Days (EHD)
See European Museum of the Year Award
See European Night of Museums
See Fête de la Science, Rennes, France
See La Nuit des Profs (Teachers' Night)
See Onderzoeksdagen (Research Days)
See Patrimoine Scientifique et Technique
   Contemporain (PATSTEC)
See QUESACO – QUEstioning Scientific
   Collections
See School team for science experiments
See Science Fair, Athens, Greece
See University of Massachusetts Boston
   Honors Program
See WND Conference 2019,
   Noordwijkerhout, Netherlands
educator   vii, 3–4, 6, 12, 21, 31, 75, 83, 100,
   159, 184
Egypt   57
Egyptian   57, 130
Einarson, Stefan   vii
Einstein and Eddington (film)   203
Einstein, Albert   203, 247–248, 250, 253,
   256–257, 260–261, 278
electric   x, 40, 44, 47, 65, 69, 75–76, 83,
   100, 103, 105, 113–114, 142, 186–187,
   191, 195–196, 228–232, 234–237, 241,
   253, 257
   arc   x, 44, 47, 113–114
   bells   195
   carriage   65, 69, 75, 83
   circuit   191, 232–233
   current   100, 257
   discharge   103, 186
   kiss   196
   motor   75–76, 83, 145
   toys   231
electrical   4–5, 36–38, 68, 75, 78–79, 98–99,
   136, 186, 195, 197, 226–231, 233–236, 242,
   247–248, 253–254, 257–258, 260, 277
   apparatus   38
   boxer   228–229
   hailstorm   227
   machine, electrical generator   37, 68,
      227, 260

INDEX

salon 4, 226–227, 230, 233–236, 242
shock 4, 230, 254
electricity viii, x, 5, 40, 67–68, 75, 79, 95, 142, 153, 181, 186–188, 195–198, 203, 206, 208, 227, 230–233, 247–248, 250–251, 253–255, 257, 259, 261, 263, 277
electro-acoustic 44
electrochemical battery 44
electrodynamics 7, 13, 89
electromagnet 69, 85
electromagnetic 9, 48, 196–198, 251–253, 258
  generator 75, 198
  radiation 251
electromagnetism 9, 48, 75, 81, 186, 196–198, 203, 250–253, 258, 263
electrometer 98–99
electron 102, 143, 151, 187, 196
electron microscope 102
electronic 77–78, 145–146, 209, 223, 273–274
  calculators 209, 223
  circuit 145–146
electronics 78, 85, 136
electrophorus 227, 233, 257
Electropolis Museum, Mulhouse, France 48
electroscope 98–99
electrostatic 1–2, 4, 37, 44, 47, 67, 69, 134, 139, 186, 188, 195, 207, 227–229, 240, 245–248, 250–254, 260, 277
  generator 2, 5, 44, 139, 186, 188, 195, 227–229, 240, 245–248, 250–254, 260
  generator (Ongemen Groote Electrizeer-Machine) 1, 245, 250–252
  machine 37, 47, 67, 253
  plate machine 251, 253
electrostatics 7, 81, 142
ellipse 219
Ellis, Robert A. 61
embodied interpretation 25
empirical practice 27
encyclopaedia x, 22–23, 123–124
enfranchisement 59
engage 2–4, 10–12, 16, 19, 30–32, 51, 59, 61, 70, 72, 75, 77, 81, 83, 103, 119–120, 158, 160, 163, 167, 169, 175, 179, 186, 189, 199, 206–207, 223–224, 243, 261, 264
Engel, Martin 242
Engels, Wolfgang 36–37, 252

English 7, 46–47, 52, 73, 123, 130, 246, 253, 271, 276
entertainment 95, 119, 203, 227, 231
entrepreneur 69, 153
environment 11, 16, 19, 26, 32–33, 60, 75, 81, 91, 95, 126, 151, 181, 190, 203, 247, 257
epistemic culture 28
epistemology 28, 88, 213–216, 220, 222–224
equal time intervals 270, 273
escape-room 181–182, 201–203, 207–208
Estonian 50–51, 53–54, 56–57, 62–63
Estonian Academy of Arts, Tallinn, Estonia 51
Estonian history of knowledge 50, 53, 57, 62–63
Estonian National Museum, Tartu, Estonia 56
ethnicity 1
ethnography 61, 139
Euclid 217
  Elements 217
Euclidean geometry 278
Euler, Leonhard 125
Europa-Universität Flensburg, Germany 228–229, 234–235, 238–239
Europe 12, 70, 244
European Day of University Collections 96–97
European Heritage Days (EHD) 127
European Museum of the Year Award 260
European Night of Museums 96–97
Eurosap-Deyrolle, Paris, France 89
evaluation 3, 6–8, 32, 57, 102, 114, 136–137, 157, 190, 224, 226, 234–236, 240, 242, 260–261
exactness 215–217
exhibition vii, 4, 7, 47, 54, 58, 62, 64, 68–75, 79, 82–83, 89–93, 101–102, 116, 126, 130, 132–133, 135, 138, 141, 152, 190, 205, 210–211, 227–229, 234–235, 264, 269
expectation 16, 57
experientialism 32
Experiment on a Bird in the Air Pump (painting) 230
experimental physics 41, 109, 119, 140, 142
experimental results 85, 141, 206, 252, 257
experimentation 30, 103, 182, 184, 189, 227, 232

experimenting   x, 2, 6, 9–10, 42, 44, 96–97, 159, 166, 241, 258–259

expert   viii, 2, 14–15, 92–93, 189, 260
  knowledge   14–15
  laboratory technician   15
  maker   15
  operator   15

explore, exploration   3, 9–10, 21, 37, 94, 110, 113, 118, 158–167, 169, 171, 173–175, 177, 179–180, 182, 187, 189–190, 202–203, 206–207, 214–215, 226–227, 267

explosion   181, 228–229

exponential   220–221, 223

extracurricular   6, 185–186, 190, 198–199, 204, 206

extracurricular laboratory   185, 199, 204

eyepiece   15, 21–24, 43, 45

Fablab, Rennes, France   90

Fabry-Buisson micro-photometer   87

Faculty of Sciences of Rennes, Rennes, France   84, 88–89

fail   4, 103, 174, 191, 194, 206, 274

Faisant, Alain   92–93

falling body   187, 267, 270, 272, 274, 277–278

Faraday cage   203, 253

Faraday effect   x, 48–49

Faraday's ring   254, 258

Faraday, Michael,   x, 48–49, 89, 196, 203, 253–254, 257–258

faux rabbit leg, See rabbit faux leg

feedback   4, 51–54, 57–60, 62, 91, 95, 136, 175, 209, 222, 224, 233–236, 248, 257, 259–260

Ferguson, James   168

Feringa, Ben   65

festival   vii, 96–97, 181, 183, 202, 207

Fête de la Science, Rennes, France   96–97

field effect transistor (FET) switch   145

field trip   158

film   viii, x, 7–8, 34–39, 41–49, 123–124, 143, 146, 148, 150, 191, 196, 203, 276

fils d'Émile Deyrolle, Les Paris, France   89

Finkener   115

fire alarm   19

firm   78
  See Bell and Siemens
  See Eurosap-Deyrolle, Paris France
  See fils d'Émile Deyrolle, Les

  See Haberdashery
  See Hartmann & Braun, Frankfurt, Germany
  See Hastavideo company
  See HistEx GmbH, Oldenburg, Germany
  See Leppin & Masche
  See Mathieu & Gentile Collin, Paris, France
  See Phywe
  See Sperry Gyroscope Company
  See Technolab, Leiden, Netherlands
  See Zimmermann, E. (Leipzig – Berlin)

Fischer, Stéphane   4, 263–279

fishing weights   111

Fizeau, Armand Hippolyte Louis   31, 123

flat-facing   24

Fleming, E. McClung   28–29

Florence   viii, 8, 34, 47, 96–99, 158, 171, 266–267, 269

fluid   70, 79, 174, 186
  mechanics   70, 174, 186
  preparation   79

fluorescence lines   9, 155

fluoride glasses   88

fluorine-related compounds   152

flying machines   203

focus   6, 14, 82, 89, 108, 161, 185, 206, 213, 219, 221, 223, 243

Fondazione Scienza e Tecnica, Florence, Italy   7, 34, 38, 47–48, 96–97

foreign exchange   130

foreign language   130

forensic   38

form-matter insight   19

formal education   1–2, 11, 237

fossil   79, 243–244

Foucault, Léon   85, 92–93, 123

four-flame gas burner   115

four-stroke cycle   76

Fourier transform infrared spectroscopy (FTIR)   151

framework   63, 101, 150, 153, 185–186, 188–189, 196, 224

France   vii–viii, 8, 10, 48, 84, 103, 122, 128, 130–131, 136–137, 146, 209, 225

Franklin hand boiler   126

Franklin's bells apparatus   186, 195

Franklin, Benjamin   187

free agency   33

INDEX 289

free fall 117, 132–133, 136, 266
   hammer and feather 276
   steel ball and feather 274
free fall demonstration 117
Free University of Brussels (VUB), Brussels,
   Belgium 141, 154
freedom 33, 214, 220, 222, 249
French 5, 10, 56, 67, 89, 122–123, 125–126,
   128–133, 146–147, 181, 205, 215, 219, 223,
   225, 265–266
French Ministry of Culture 126
French school of 10, 181
frequency generator 145
frictionless track 70
Friends of the historical museum of the
   Hoche high school, *See* amis du musée
   historique du Lycée Hoche, Les
Frize, Bérengère 88
frogs' heart 73
frustration 30, 36
functionality 17, 19, 24, 156, 162, 249,
   252–253

Galilei, Galileo viii, 4, 31, 34, 123, 167,
   171–174, 179, 182, 186–187, 190–191, 201,
   263, 265–273, 275–278
   Dialogue Concerning the Two Chief World
      Systems 266
   Discourses Concerning Two New
      Sciences 187, 265–266, 268–269, 273,
      275
Galileo (geometrical) compass viii, 167,
   171–172, 179
Galileo's pendulum 201
Galileo's telescope 186
Galinstan 255–256
Galison, Peter 184
gallery 5, 19, 85, 87, 90, 94, 96–99, 102,
   162–166, 175, 177, 186, 248
Gallon, Benjamin 89
galvanometer 2, 54, 132–133, 135, 139, 142,
   238–239, 255, 258–259
game 70, 91, 181–182, 201, 203, 275
   theatrical 203
   word 201
Ganot, Alphonse 39, 270
Garden Observatorium, Teylers Museum,
   Haarlem, Netherlands 247
gas burner 115

gas light 52, 54–55
gauge 38
Gauvin, Jean-François 173, 180
Geissler's tube 38, 44
Genaille's rod 210–211
gender 1, 12, 234–235
   identity 1
genetics 264
Geneva 4, 9, 263–265, 269
Geneva Museum of the History of Science,
   Geneva, Switzerland 4, 9, 263
geography 3, 12, 48, 60, 125–126
geology 50, 54, 95, 98–99
geomatics 153
geometric 10, 24, 29, 171, 209, 215–219, 223
geometry 158, 171–173, 175–176, 178–179,
   215–220, 223, 270, 278
German x, 50, 52, 55–56, 77–78, 106–107,
   130, 195, 237
Ghent University Museum (GUM), Ghent,
   Belgium 8–9, 139, 148, 150, 152, 154, 156
   Collection for the History of Sciences,
      Ghent, Belgium 154
Giatti, Anna 35, 38, 43, 45–46
Gibson, James 16–17
Gigault, Christian 146
Gilborn, Craig 28
Giordano, Ray 163
Gires, Francis 123
Gizelis, Dimitrios 106–107
Glass and Ceramics Laboratory of Rennes,
   Rennes, France 87
glass rod 1, 164, 166
global warming 264
Goble, Mickael 87
God 68
Goethe 240
Gomez, Carolina 165, 180
Goodyear, Peter 61
Google-form 57
Gorredijk, Netherlands 76
Graetzin-Licht gas light 55
gramophone 5, 95
Grandjean, Daniel 88
graphic design 50–51, 53, 57–58, 63–64, 72
graspable 16
Gravesande, Willem Jacob 's
   Physices elementa mathematica 70–71
Gravesande, Willem Jacob 's 70–71

290 INDEX

gravimeter 132–133
Greece vii, 6, 106–107, 207, 216
Greek 6, 20, 105–107, 109, 182, 185, 188–189, 191, 199, 217
Grenet electric cell 113, 254
Groningen Physics Society, Groningen, Netherlands 69
Gu, Mingwei 163–164, 166, 179–180
Guan, Charles 180
Guericke, Otto von 182, 187–188, 192
guided tour, *See* tour
Guillemin, Amédée 39–40, 45, 270
gunpowder 68, 228–229
Guy, Graziella 89
gyroscope 85, 160

H.M. the Queen of the Netherlands 72
Haarlem Municipal Theatre, Haarlem, Netherlands 247
Haarlem, Netherlands 1, 48, 81, 98–99, 243, 245–247, 250, 252
Haberdashery 73
Hahn, Otto 204
handle viii, 9, 11, 16, 19–23, 28, 60, 72, 79, 94, 156, 162, 179, 206–207
hands-on 1, 4, 11, 24, 50, 57–59, 66, 78, 80, 167, 170, 174
hands-on activity 1, 11, 58, 78, 80, 167, 170
Hannan, Leonie 51, 58–60, 62
haptic 28–29
Hartmann & Braun, Frankfurt, Germany 108
Harvard University, Cambridge MA viii, 28, 157, 160, 173, 187
The Collection of Historical Scientific Instruments, Cambridge, MA 120, 172–173, 180, 263
Hastavideo company 35
Hauksbee, Francis the Elder 67
Haüy, René Just 270
Traité élémentaire de physique 39, 270
Havo- higher general secondary 248, 261
health and safety officers 73
health regulations 252
Heering, Peter vii, 1–13, 64, 180, 208, 226–242, 268
Heisser, Ronald 174, 180
Hellenic Foundation for Research and Innovation 120

Helsen, Josef 204
Cahier de Ma Vie en Prison 204
Merveilles de la Science, Les (sketches) 204
Hémery, Corentin 219
Heritage Days, *See* European Heritage Days (EHD)
Heron's fountain 85–86, 94–95, 150
Herschel, William 151
Hertz experiment 48
Hervault, Valentin 87
Heubel, Friedrich 57
Heymans Institute for Psychological Research, Groningen, Netherlands 77, 82–83
Heymans, Gerard 77, 83
hierarchy 10
high school, *See* school
high-voltage 43, 68, 134
HistEx GmbH, Oldenburg, Germany 253
historical
    artefact 6, 36, 38, 48, 213, 226
    context 1, 78, 149–150, 154, 166, 230, 233, 277–278
    physics 6, 122–123, 125, 127, 129, 131–133, 135–137, 247
    rooms 182
historiographic 28–31, 184
history of concepts 279
history of ideas 263
history of music 30
History of Science and Medicine Library, Brill vii
history of techniques 279
history of the museum 203
history of touch 59
Hittorf's tube 38
Hogere Burger School, Groningen, Netherlands 69
Holmes, Sherlock 7, 181–182, 203
Holtz induction machine 256–258
Hooke, Robert 202
Houk, Peter 164, 180
House for Science in Brittany at the service of teachers, *See* Maison pour la science en Bretagne au service des professeurs, Rennes, France
Huisman, J.W. 4, 65–83
human preparation 79

INDEX 291

humanities 10–11, 50–51, 57, 61, 68, 82
Hunt, J.L. 146
Huygens 154
Huygens vacuum pump 154
hydrogen balloons 199
hydrostatic balance 150
hyperbola 219–220
hypothetical compass 217

imagination 3, 14, 18–20, 30–31, 37, 44, 91,
    95, 165–166, 181, 190, 197–198, 204, 220,
    223, 268
Imitation Game, The (film) 203
Imperial College London, UK 72–73
implementation 59, 96–97, 115, 117, 120, 190,
    220, 237, 240, 257, 260
in-house training 51
inclined 4, 8–9, 191, 265–274, 276–279
    plane 4, 8–9, 191, 265–267, 269–274,
        276–279
    rope 269
indigenous science 12
induction 44, 196–197, 233, 256, 258, 277
    electromagnetic 196–197, 258
    electrostatic 233
    mutual and self 197
induction coil 44, 196, 277
    Ruhmkorff 186, 188, 249, 256, 258
inert gas 182
infinitesimal calculus 218
Information System of Estonian Museums
    54
    Museums Public Portal 54
Inhelder, Bärbel 9, 160
inquiry 33, 60–61, 172, 189, 196, 206, 257,
    260–261
inquiry-based 61, 257, 260–261
inquiry-based assignment 61
Institute for Research in the Teaching
    of Mathematics (IREM), Brest,
    France 209–212, 224
Institute of History and Archaeology, Tartu,
    Estonia 10, 50, 52
instruction 2, 8, 11, 17, 19, 103, 111, 128, 130,
    140, 154, 159, 190, 196, 201, 212, 227, 230,
    240–241, 258–259, 261
instrument maker 55, 66, 69, 136, 139, 146,
    163, 251
intellectual culture 31

intellectual pleasure 13
interactive museum 36, 174, 182
intern 62, 85–89, 94, 104, 132–133, 139
internal combustion 69, 76
internal combustion engine 69, 76
International Union of History
    and Philosophy of Science and
    Technology vii, 3
International Year of Crystallography 89
International Year of the United Nations 90
internet 3, 84, 90, 95, 111, 114
interpretation 6–7, 9, 21–23, 25–26, 33, 47,
    100, 170, 230, 241
interrupter 38, 145
interview 4, 54, 88, 102, 233, 236, 240–241
inventory 11, 54, 85, 87–89, 151
inverse tangent 218, 220
investigation 6, 9, 57, 103, 118, 134, 154, 159,
    161, 174, 180, 189
Invisible Man, The (film) 203
ionisation chamber 98–99
iPad 8, 150
IR-UV spectrophotometry 151
Islamic science 174–175
ISO 53
isochronism 201, 269
isolated sensors 16
Istituto Galileo Galilei, Florence, Italy 34
Istituto Tecnico Toscano, Florence, Italy 7,
    34
Italian 7, 34, 46–47, 130, 187
iteration 21, 171, 197
itinerant lecturer 227, 232, 240

Jacob's staff 224
Jacomy, Bruno 123
JavaScript coding 178
Jobin 87
Johnston, Steven ix
Joliot, Pierre 98–99
Jones, Indiana 167
Joubin, Louis 89
Jouet, Olympe 86
Juicy Salif x, 18–19
jury 91, 157

kaleidoscope 182
Kamprani, Katerina x, 19–20
Karm, Mari 64

Keil, Inge   vii
Keiss, Aleksander Paul   55
Kepler's telescope   186
Kerpan, John   165, 180
Khantine-Langlois, Françoise   5, 8, 122–138
Kiley, Sharon   165, 180
kindergarten, *See* school
Kirchhoff, Gustav   188
Knallgas   76
know-how   41
knowledge of scientific instruments   14
Koenig acoustic siren   44
Koenig, Rudolph   44, 85, 91–93
Koffi, Gerald   165, 180
Kohl, Max   115, 249
Krippendorf, Klaus   17
Kuhfeld, Ellen   228–229
Kwan, Alistair   11, 14–33, 176
kymograph   1, 4, 72–75, 77, 82

La Nuit des Profs (Teachers' Night)   95
Laboratorio delle Macchine Matematiche, Modena, Italy   213
laboratory assistant   41
Laboratory of Laser Physics, Rennes, France   87
Laboratory of Physics of Infra-Red Spectroscopy, Rennes, France   87
Laboratory of Physics, Haarlem, Netherlands   249
laboratory project   199, 204
laboratory technical assistant   105
LabView   145–146
Laidla, Janet   10–11, 50–64
lamp   38, 42–43, 46, 55, 66, 111, 113–115, 123–124, 151, 164, 191
lamp of Locatelli   151
Landau, Eber   x, 55–56
Landgrave Charles of Hesse-Kassel   37
Langevin, Hélène   98–99
Largaespada, Raul   169–170, 180
laser   87, 167, 171–175, 178–180
laser-cut   167, 171–175, 178–179
Latin   12, 33
Latin America   12
Lauginie, Pierre   5, 122–138
Lavoisier, Antoine-Laurent de   125
Lavoisier calorimeter   94

law   vii–viii, 31, 39, 126, 150, 257, 264, 267, 270, 276–278
   of falling bodies   267, 270, 277–278
   of universal gravitation   276
Lazos, Panagiotis   6, 105–121
Le Clanche, Julien   87
lead (metal, element)   38, 268, 272
Leaning Tower, Pisa, Italy   31
learned institution
   See Academy of Sciences, France
   See Accademia del Cimento, Florence
   See Association Science en Seine et Patrimoine (ASSP), Rouen, France
   See Centre François Viète, Brest, France
   See Fondazione Scienza e Tecnica, Florence, Italy
   See International Union of History and Philosophy of Science and Technology
   See Laboratory of Physics, Haarlem, Netherlands
   See Maison pour la science en Bretagne au service des professeurs
   See Rennes en Sciences Association, Rennes, France
   See Royal Institution, Great Britain
   See Royal Society, London, UK
   See Société Française de Physique, France
learning   1, 8, 11, 15, 17, 31–33, 51, 58–64, 68, 80, 103, 116, 119, 132–133, 158, 161, 169–170, 180, 214, 222, 241, 243, 247, 257, 261
   case-based   61
   inquiry-based   61, 257, 260–261
   object-based   51, 58–63
   object-centred   50, 60
   object-inspired   64
   problem-based   61
   project-based   61
Lebedev, Ilia   180
Lebossé, Camille   219
Leclanché electric cell   113
lecture   190, 196, 199
LED lamp   191
Leeuwenhoek, Antoni van   9, 143, 154–156, 163
Lefeuvre, Marie-Aude   89
Leibniz, Gottfried Wilhelm   218
Lemaire, Marion   104

INDEX

lens   6, 9, 24, 45, 143, 155, 158–159, 163, 166, 169, 188–189, 237, 240
Leppin & Masche, Berlin, Germany   108
Leslie cube   151
Leupold, Jacob   37
Leybold's Nachfolger vacuum pump   81
Leyden jar   47, 68, 111, 139, 142, 186, 195–196, 199, 207, 227–232, 246
Li, Tongji   172, 179–180
Liceo Paolo Sarpi, Bergamo, Italy   8, 34, 47
Lichtenberg figures   227, 233
light-box   53
lightning   68, 195, 228–229, 233
   conductor   68, 228–229
Likert scale   136, 233–235
limelight   123–124
Limelight (film 1952)   124
linear   217, 241
Linkberg, Artur   54
Lipnoski, Sandi   180
Lissajous curves   197–198
Lissajous, Jules Antoine   129, 197–198
literature   17, 51, 62, 64, 130, 152, 184, 189, 206
Liuni, Francesca   174–178, 180
live-wire   vii
Loewi, Otto   73
logarithmic   209, 218–220
   curve   218–220
   functions   209
London, UK   3, 13, 58, 64, 72–73, 77, 182, 207
Lord, *See* God
Lorentz Lab, Teylers Museum, Haarlem, Netherlands   13, 81, 243–245, 247–262
Lorentz, Hendrik Antoon   245–247, 250, 253, 260–261
Louis XV, King of France   128
Louis XVI, King of France   128
Louis XVIII, King of France   128
Lourenço, Marta   123
Lucas, Jacques   92–93, 104
luminescence   67
lute   267, 269
Lycée Hoche, Versailles, France   125–126, 128–133

machine for reproducing the polar aurora   279
machinist   29

Magdeburg   182–183, 188, 192, 264
   hemispheres   192
magnetic needle   197–198
magnetometer   102
   Superconducting Quantum Interference Device (SQUID)   102
Main à la Pâte Foundation   94
mains electricity   79
Maison pour la science en Bretagne au service des professeurs, Rennes, France   94–95, 98–99, 104
maker   6, 15, 35, 42, 46, 52–57, 60, 62, 66, 69, 80, 116, 136, 139, 146, 160, 163, 165–166, 174, 251
maker's logo   60
Maltese Falcon (ship)   167
manometer   38
Mantes-la-Jolie, France   130
Mantua, Duke of   172
manual   16, 114, 130, 167–169, 171–172, 176, 197, 241, 251, 253, 260, 277
manufacture   77, 79, 89, 102, 106–109, 115, 143, 155, 224
manuscript   8, 75, 161, 165–167, 169–170, 174–177, 179, 267
Maraslean Pedagogical Academy, Athens, Greece   106–108
   Laboratory of Natural Sciences   108
Maraslean project   119–120
Maraslean Teaching Center, Athens, Greece   6, 105–108, 120
Maraslis, Grigorios   106–107
Marey's recording cylinder   95–97
Martin, Philip   203
Marum, Martinus van   2, 37, 245–246, 249–252, 254, 260
mass spectrometer   102
Massachusetts Institute of Technology, (MIT) Cambridge MA   viii, 9–10, 152, 158, 160–170, 172, 174–177, 179–180
   MIT Edgerton Center   9, 158, 161, 171, 180
   MIT Glass Lab   180
   MIT Museum   10, 162–164, 167–170, 180
   MIT Wallace Astrophysical Observatory   170
master's degree   8, 86–89, 118, 122, 132–135, 139, 141, 150–151, 154, 157, 175–176, 212–214
matches (fire ignition)   43

material culture 14–15, 28–29, 59, 112,
    183–185
materiality 2, 30, 213, 222
materials science 140, 142
Mathematical-Machines Laboratory, *See*
    Laboratorio delle Macchine Matematiche,
    Modena, Italy
mathematics vii, 10, 50, 65, 69, 120, 140,
    146, 178, 209–211, 213–215, 217–218,
    223–225
Mathieu & Gentile Collin, Paris, France
    56
McCarthy, Brian 161, 180
McCracken, Grant 28
meaning-making 19, 60
measurement 24, 84, 86, 94, 99, 101–102,
    123, 126, 128, 136, 141, 146, 151, 160, 179,
    192, 196, 198, 206, 259, 266, 273, 278
mechanical 17, 36, 78, 96–97, 103, 136, 144,
    146, 173, 212, 218, 223–224, 259
    calculator 96–97, 212, 224
    engineering 173
    gear transmission 144
    models 36
mechanics 7, 136, 186, 264, 266
median 126, 234–235
Medical Research Council (British) 72
medicine vii, 4, 27, 50, 54, 61, 72–73, 79,
    82–83, 139, 187, 233, 245
Meinnel, Jean 88
Melloni, Macedonio 150–152, 154
Melloni bench 150
mercury viii, 38, 42, 66, 116–117, 136, 177–178,
    182, 203, 251–252, 255–256
    fine shower demonstration 116
meridian transit 24
Merveilles de la Science, Les (sketches) *See*
    Helsen, Josef
metaphor 28, 184
meteorite 79
metronome 273–274, 278
Michelson, Albert A. 160, 162
microbalance 101
microscope 9, 31, 45, 102, 139, 143, 151,
    154–156, 158, 163–166
middle school, *See* school
Mikelsaar, Raik-Hiio 54
Milici, Pietro 9–10, 209–225
Milioni, Vasiliki 182, 208

military specifications 77
minerology 57
Miniati, Mara ix
Ministry of Education, France 131, 138, 190
mirror 143, 146, 148, 161–162
MIT, *See* Massachusetts Institute of
    Technology, (MIT) Cambridge MA
Mittelbau-Dora concentration camp 205
model 2, 8, 10, 24, 26, 36, 38, 57, 61–63, 68,
    70–71, 73, 75–76, 94, 130, 149–150, 163,
    167, 171–175, 179, 195, 219, 228, 269
Modern Times (film) 203
Mohs hardness test 57
Montessori pedagogical tools 213
Moon 276
Moreau, Georges 88
Morgenstern, Johann Karl Simon 57
Moriarty, Edward 171, 180
Moriarty, Peter 180
Morin free fall machine 132–133, 136
Morin's apparatus 270, 272
morphology 139
Morrison, James 171
Morrison, Philip 171, 180
Morrison-Low, Alison viii, 13
Morse electric telegraph 105
moving images 143, 146
Munroe, C.J. 173, 180
musaeum 244
Musée Curie, Paris, France 98–99
Musée des arts et métiers (MAM), Paris,
    France 210–211
Museo Galileo, Florence, Italy viii, 171–172,
    269
museum vii–viii, x, 1–5, 7–11, 13–23, 30, 32,
    36–37, 52–56, 59, 61, 63, 65, 72, 75–76,
    78–79, 81–83, 98–99, 102, 126, 128–130,
    139–143, 150–157, 162–163, 170, 174–175,
    179, 182, 186, 199, 201, 203–204, 207, 212,
    233, 243–248, 252, 258, 260–264, 269,
    272–274, 278–279
    education 14, 32, 51
    ethics 83
    university x, 3, 6, 8, 50–51, 56, 59, 67,
        69–70, 73, 139
    See Athens University History Museum,
        Athens, Greece
    See Deutsches Museum, Munich,
        Germany

INDEX

See Electropolis Museum, Mulhouse, France

See Estonian National Museum, Tartu, Estonia

See Geneva Museum of the History of Science, Geneva, Switzerland

See Ghent University Museum (GUM), Ghent, Belgium

See interactive museum

See MIT Museum, Cambridge MA

See Museo Galileo, Florence, Italy

See Museum for the History of Sciences, Ghent, Belgium

See Museum Mensch und Natur, Oldenburg, Germany

See Museum of Geoastrophysics of the National Observatory, Athens, Greece

See Musée Curie, Paris, France

See Musée des arts et métiers (MAM), Paris, France

See Natural History Museum of Geneva, Geneva, Switzerland

See Phänomenta Flensburg, Germany

See Teylers Museum, Haarlem, Netherlands

See University Museum Groningen (UMG), Groningen, Netherlands

See University of Tartu Art Museum, Tartu, Estonia

Museum for the History of Sciences, Ghent, Belgium 139, 154

Museum Information System 62

Museum Mensch und Natur, Oldenburg, Germany 227

Museum of Geoastrophysics of the National Observatory, Athens, Greece 117

Museum of the University Groningen, Netherlands 4, 65

music record 96–97

Musschenbroek, Jan van 37, 66

Musschenbroek, Pieter van 187, 196, 240

mystery 6–7, 203

Napier's rod 210–211

Napoleon 37, 126, 244

narration 1, 4, 9, 29, 199, 201, 233

National School of Chemistry, Rennes, France 101

Natural History Museum of Geneva, Geneva, Switzerland 269

natural magic 182, 188, 202

nature of history of science 52

nature of science 14, 159, 166, 241, 258–259

Natzweiler 205

navigation 22–23, 125, 182, 202–203, 212

Netherlands vii, 1, 65, 69, 72, 76, 243, 252

network 1, 44, 92–93, 96–97, 102, 104, 116, 137–138

neuro-technologist 73

neurology 55

neusis 217

Newton pendulum 112

Newton's experiment 40

Newton's tube 270–271

Newton, Isaac 40, 70, 112, 182, 187–188, 202, 270–271, 276–277

nickel tubes 88

Nickel, B.G. 146

Nicomedes, *See* conchoid

Night and Fog Decree 205

Nikolaidis, Efthymios 108

Nobel laureate 65

Nobel Prize 73, 83, 98–99, 245

Nollet's double cone 186

Nollet, Jean-Antoine 128, 130, 187, 232, 269

Leçons de physique expérimentale 269

non-algebraic 223

non-university higher education 140–141

Nordmetall-Stiftung 242

North America 12

NOS, *See* nature of science

novice 2, 29, 226

nuclear energy 100

nuclear model 24

Obellianne device 89

object handling 11, 50–51, 53, 57–59, 61–64

object's biography 61

object-based 51, 58–63

object-based research 59, 61, 63

object-centred 50, 60

object-inspired 64

observation 9, 11–12, 16, 31, 59–60, 90, 103, 158, 166, 179, 187–188, 191, 240, 259, 266–268, 273, 276

stations 16

observatory 36, 117, 170, 260
octant x, 21–23
Odessa, Ukraine 106–107
office phone 52, 54, 56
Ohm's law 126
oil diffusion pump 101
Ombrédanne, Louis 56
Onderzoeksdagen (Research Days) 75
Ongemeen Groote Elektrizeer-Machine 1, 245, 251–252
open questions 137
open-ended 115, 158
operation viii, 2, 7, 11, 16, 36, 75, 83, 86, 94, 98–99, 103, 116, 190, 196, 217, 237, 251–254, 260, 264, 278
operator 15–17, 22–24, 26, 31, 38, 46
optical level 161, 169–170, 179
optics 7, 22–24, 28, 39–40, 43, 48, 142–143, 155, 158, 163, 169–170, 179, 186, 263
oral 58, 154, 233
  feedback 58
  presentation 154
orrery 36, 279
oscillatory fall gravimeter 132–133
oscilloscope 129, 196–197
outreach 132–133, 137, 237, 244
Oval Room 245–247, 251–252
Overgauw, René 254
ownership 59–60, 153, 240
oxyhydrogen 76

Padua 173, 266, 275
Pal, George 203
palaeography 33
Panusch, Martin 236, 242
Paolo Sarpi High School, Bergamo, Italy 8, 34, 47
Paparou, Flora 6–7, 181–208
Pappus of Alexandria 217
parabola 10, 136, 219–221, 223
parabolograph 220–221
parents 6, 108, 127, 129, 167, 180
Paris x, 56, 99, 108, 132, 181, 218
Park, Yoonah 180
Pascal's law 150
Patrimoine Scientifique et Technique
  Contemporain (PATSTEC) 85, 87, 101–102

PATSTEC, See Patrimoine Scientifique et
  Technique Contemporain
pedagogy 8, 14, 28, 32, 59, 61, 63, 90, 103, 105–109, 122–123, 130, 136–138, 159, 170, 209, 213–215, 222–223
Pena, Octavio 102
pendulum 36, 77, 92–93, 111–113, 159, 171, 195, 201, 265–266, 268–269, 275–276, 278
  bifilar 268, 275
  isochronism 201, 269
  period 268–269, 275–276
  reference 275
Perani Serra, Laura 34
performance viii, 7, 30, 48, 142, 181–182, 187, 189, 199, 204, 247, 252–253, 260
performer 30, 43, 182, 207
perfume sprayer 70
Perks' instrument for the quadrature of the
  hyperbola 220
perpendicular 24, 221
Perrault's construction 218
Perrault, Claude 218
Perrin, André 90, 101, 104
Perrin, Christiane 90, 101, 104
persistence of vision 143, 146
personal 3–4, 10–11, 21, 31, 37, 51, 58, 68, 118, 137, 159, 163, 166–168, 170–171, 174, 177, 179, 199, 224, 244, 259–260
  equation 31
  experience 10, 51, 166–167, 170, 174
perspective-taking 175, 179
petanque ball 274
Petit, Axel 87–88, 146
Phänomenta Flensburg, Germany 4, 226–227, 230, 232, 234–237, 242
pharmacy 50, 56, 73
phenakistiscope 146
phenomena viii, 1, 4–5, 7, 17, 66–70, 100, 103, 126, 146, 179, 186, 188–189, 231, 233, 257–258, 260, 267, 274–275, 278
philosophical amusement 24
philosophy vii, 3, 9, 24, 29, 108, 141, 146, 159, 175, 184–185, 187–189, 192, 194, 207–208, 213, 244, 267, 271
  of experiment 184
  of science vii, 3, 108, 184–185, 188–189, 213, 267
phonograph 105, 113, 116

INDEX 297

phosphorescent screen 24
photo-chemistry 152
photographic enlarger 91
photography 11, 50–51, 53, 59, 62, 163
physical context 75, 230
physicist 3, 6–8, 85, 87–89, 129, 245
physico-theological 68
physics cabinet 8, 36, 43, 46, 66, 123, 128–129, 143, 150, 270
physiology 73, 146
Phywe 106–107
Piaget, Jean 9–10, 160, 214
piano 19, 204
Pigelet, Bernard 98–99
Pisa 266, 269
Plaine, Jean 89
planar 217
planetarium 69
planimeter 210–211, 219
Plantevin, Frédérique 9–10, 209–225
Plateau, Joseph 8, 139, 142–146, 148, 150
Plato 14, 19, 31
Plato's cave 31
Plato-Aristotle complementarity 19
play (theatrical) 181, 201, 207, 264
plumb line 172
pneumatic 37
    pump 37
pneumatics 187, 263
pocket watch 218
pointwise 219, 223
Polanyi, Michael 29–30
polarimeter 45
polarisation 152
polynomial equations 217
popularisation of science 103
possibility 2, 11, 13, 16, 37, 68, 83, 100, 132–133, 138–140, 145, 150, 159, 175, 191, 193, 218, 222, 253
postcard 123, 125, 128–129
poster 11, 52–58, 60–62, 64, 90–91, 98–99, 132–133, 174–175, 210–211, 230–231, 233
postgraduate students 109, 112, 115–117, 119–121
power supply 254, 256
pre-defined experiments 78
préparateurs (laboratory assistants) 41
preservation 8, 39, 108, 118, 120–122, 137–138
primary, *See* school

principle of equivalence 278
Priser, Julie 5, 7, 84–104
prism 24, 40, 87, 188–189, 202–203, 240–241
problem-based 61
professional restorer viii
project-based 61
proportion 206, 268
proportional 100, 145, 219
    compass 219
    integral and differential (PID) controller 145–146
proprioceptive 29
protocol vii–viii, 28, 136
Prown, Jules 28–29
psychology 9, 16, 50, 52, 54–55, 60, 73, 75, 77, 80, 82–83
public institution viii
publication 3, 60, 67
pullable 19
pulley 74, 271–272
push 19
pushable 16
Puusepp, Ludvig 55
puzzle 21, 57, 182
Pyramid of Caius Cestius 57
Pyrex 164

QR code 136
quadrant 98–99, 160, 167, 173–174
quadrant electrometer 98–99
quadratrice 217
quadrature 217, 220
quartz current generator 98–99
quartz scale 98–99
Quéré, Yves 123
QUESACO – QUEstioning Scientific COllections 90–93
questionnaire 57–58, 91, 115, 118, 136, 233–236, 240, 248, 260
questions-and-answers 216
quiz 136

rabbit
    faux leg 73
    fur 1, 230
    leg 73
race 1
radiation 38, 86, 150–152, 251
radioactive source 98–99

radioactivity   viii, 7, 84, 98–100
raison d'être   263
ramp   4, 171
Rayleigh   182, 202
reading telescope   45
reboot   50, 60, 63
reciprocity   11
reconstruction   3, 8–10, 65, 76, 84, 98–99, 120, 191, 208, 215, 220–221, 226, 241, 252, 263, 269
rectifier   44
Reed, Edward   32
re-enactor   5, 183
reflection   x, 3, 9, 12, 25, 27, 103, 110, 159, 161–162, 166, 179, 278
reflector   24
regulation   viii, 228–229, 252
rehearsal   14, 32, 181
religion   67
remote-sensing satellites   16
Renaissance   12, 31, 36
    Pyrrhonists   31
Rennes en Sciences Association, Rennes, France   84, 89, 92–93, 104
Rennes Institute of Chemical Sciences, Rennes, France   90, 102
Rennes, France   5, 7, 84–94, 96–102, 104, 209
repair   2, 6, 8, 106–108, 111–113, 118–120, 150, 198
replica   2, 4–5, 9, 36–37, 60, 62, 65, 72, 76, 80, 83, 94, 142, 146, 149, 154, 156, 160–162, 215, 238–240, 242–243, 247–257, 259–264, 276–277
replicate   vii–viii, 31, 36, 69, 75, 83, 163, 184, 193, 250
replication   21, 31, 176
researcher   2–3, 13, 61, 85, 87–93, 101–102, 104, 159, 171, 184, 267
restoration   viii, 32, 38, 132–134, 138, 140, 249
restorer   vii, 48
Resurrection   67
RETE   73, 178
    archive   73
    forum   73
Ribokov Cage   80
Richard, Gaëlle   89
Ridoux, Olivier   91, 104
Riess' Spiral   254, 258
Ritchie motor   254, 258–259

rod   1–2, 24, 164, 166, 196, 210–211, 221, 230, 233
Rodriguez, Lillian   165, 180
rods   1–2, 24, 164, 166, 210–211, 230
Rome   57
rotating colour disc   77
rotating drum   73, 83
Rowe, Shawn   60
Royal Institution, Great Britain   182, 202–203, 207
Royal Society, London, UK   x, 25, 27, 181, 183, 187, 202, 207, 219
Royal Teaching School, Athens, Greece   105
Rozé, Nathalie   85–86
Ruhmkorff coil   186, 188, 207, 249, 256, 258
ruler   100, 223, 266–267, 273
Russian   52
Rutherford, Ernest   24

safeguard   38, 87, 94
safety   viii, 2, 7, 36, 38, 73, 79, 136, 201, 228–229, 251, 254
    hazards   2, 7, 38
    rule   36, 38, 136
Salviati   266–267
San-Miguel, Alfonso   5, 8, 122–138
Sanskrit   33
Sarton, George   157
Sauer, Friedhelm   242
scale   21, 46, 57, 75–76, 87, 98–99, 136, 171–172, 189, 233–235
scarificator   30
Schallschlüssel (Voice Key)   78
Schechner, Sara   ix
Scheiner's pantograph   219
Schmid, Vanessa   236, 242
Schmidt, Carl   57
Scholierenacademie   75
school   3, 5–6, 8–10, 34, 36–37, 41, 46, 51, 69, 75, 78–79, 82, 84–85, 93–97, 101, 105–109, 111–112, 116–117, 119–122, 124–131, 137–139, 140–141, 150, 153, 165, 181–183, 184–186, 194–195, 198–199, 201–204, 206–208, 213, 215, 223, 225–227, 237–240, 247–248, 250, 253, 256, 258–263
    elementary   5, 92–93, 129
    high school   3, 5–6, 9–10, 34, 41, 46, 85, 92–93, 95–97, 109, 112, 116–117, 119–120,

128–130, 139, 141, 150, 153, 181, 199, 223, 225, 263

kindergarten 213

laboratory 6, 108, 111, 121, 181–182, 185, 198, 206

middle 85, 128, 130

primary vii, 51, 94–97, 105, 119, 138, 209, 213, 227, 277

programme 95, 108, 250, 253, 261–262

secondary vii, 51, 69, 75, 78–79, 82, 122, 124–127, 130–131, 137–138, 185, 209, 226–227, 237, 240, 248, 256, 277

Technasia – pre-university 261

vmbo -lower vocational 260

vwo – pre-university 248, 261

See Corfu school

See French school

See Havo- higher general secondary

See also 1st Upper High School (Lyceum) of Vrilissia, Athens, Greece

See also 26th General Lyceum of Athens, Greece

See also 39th Primary School of Athens, Greece

See also Athens Teaching School, Athens, Greece

See also Collège du Château, Morlaix, France

See also Hogere Burger School, Groningen, Netherlands

See also Istituto Galileo Galilei, Florence, Italy

See also Istituto Tecnico Toscano, Florence, Italy

See also Lycée Hoche, Versailles, France

See also Maraslean Pedagogical Academy, Athens, Greece

See also Maraslean Teaching Center, Athens, Greece

See also Paolo Sarpi High School, Bergamo, Italy

See also Royal Teaching School, Athens, Greece

School team for science experiments 108

Schotte, Ernst (Berlin) 108

Schwarzburg, Germany 57

science

centre 2–4, 36, 226

communication 181, 183–185, 189–190, 198–199, 202–204, 206–208, 260

competition 185

education 1, 3, 8, 12, 30, 90, 103, 105, 118–121, 226, 241, 257–258, 260

Science Fair, Athens, Greece 181, 183, 202–203

Science Technology Engineering Math (STEM) 120

science- *wissenschaft* 50, 227, 245

science-and-society 203, 206

science-theatre 198, 202, 204, 206

Scientific Culture Commission, Brittany, France 104

Scientific Instrument Society (SIS) 50, 52, 84, 86, 104, 123, 146

scientific instrumentation 21, 137

Scientific Instruments and Collections vii–viii

scientific knowledge 69, 185, 238–239, 258–259

Scientific University College, Brest, France 211

score 30, 57–58

scuba tank 44

secondary, *See* school

Segers, Danny 8, 139–157

self-efficacy 13, 240

self-registering data loggers 16

self-registering instrumentation 31

serial number 60

series vii–x, 5, 14, 34, 36–38, 47, 66–67, 101, 182, 186, 190–191, 201, 230, 267, 275

set square 220

Settle, Thomas B. 191–192, 267, 275

sextant x, 11, 21–23, 31, 167, 170, 172, 175

Short, James 85

shower cubicle 70

shutter speed 53

Simou, Polyxeni 182, 208

sine galvanometer 255, 258–259

single-lens microscope 9

siphon 2

siren 77

acoustic 44, 115–116

civil defence 115

sit-on-able 17

skeuomorphic design 17

skill 11, 22–23, 29, 31, 35, 51, 53, 58, 60–61, 63–64, 91–93, 139, 141–143, 146, 164, 174, 179, 185, 208, 214, 225, 240–241, 257, 259, 261

Skordoulis, Constantine   6, 105–121
Skype   73
slide rule   5, 95–97, 209
Slikker, Jacob   254
Sly, Jackie   180
Smartphone   42, 45
smoked paper   73–74
smoked-drum kymograph   1, 4, 72
Smoot, Oliver   169, 179
social sciences   82
Société Centrale de Produits Chimiques   87
Société Française de Physique, France   132–133, 138
Söderlund, Inga Elmqvist   vii
solar microscope   31
solar spectrum   40
solid   88, 217
solid-state chemistry   88
sound analyser   91–93
soundtrack   8
source   14, 17, 21, 29, 43–44, 46, 51, 59, 103, 123, 207, 227
    primary   21
    three-dimensional   51
source document   21
spark plugs   76
sparks   5, 195, 198–199, 231, 245, 252, 277
spectacular   7, 42, 192, 203, 261, 263–264, 277
spectra   45, 48, 188
spectroscope   x, 14, 44–46, 87, 189
spectroscopy   43, 87, 151, 182, 186, 188, 201
spectrum   40, 46, 182, 203
speed of light   31, 123, 160, 162
    experiment   160, 162
Spek, Trienke van der   4–5, 81, 243–262
Sperry Gyroscope Company   160
Sperry mirror unit   161
sphere   86, 154, 163–164, 175, 266–267
    See armillary sphere
    See celestial sphere
spinthariscope   x, 1–2, 24–25, 27, 31
spiral   217, 254, 258
Srikanth, Rajini   180
staff   105, 230, 237
Starck, Philippe   x, 18–19
static electricity   68, 187, 195–196, 198, 250, 257
statistical analysis   77
statistics   77, 236

steam   36, 38, 69, 76, 94, 148
    engine   36, 38, 94, 148
    engine model   38
    wagon   69, 76
steampunk   14
steel tape (surveying)   169
Stefanidou, Constantina   6, 105–121
STEM, *See* Science Technology Engineering Math
step-on-able   16–17
stepladder   274
stereoscope   2, 111, 113, 116, 146, 148–149
stereoscopic photograph   111, 146, 148–149
Stern-Gerlach   31
Stilwell, Gary   167, 175–178, 180
Stockholm Junior Water Prize   130
stockpot   20
Stoelwinder, Anton   76, 254–255
stopwatch   4, 272–273, 278
storytelling   248
straightedge   217
    unmarked   217
Strano, Giorgio   ix, 13, 208
Stratingh, Sibrandus   65, 69, 75–76, 81, 83
strobe   160–161
Strutt, John William, 3rd Baron Rayleigh   182, 202
student organization
    See Association des Physiciens de Lyon (APL)
    See Groningen Physics Society, Groningen, Netherlands
Studium Generale (Groningen)   83
subtangent   220–223
Sun   86, 160, 175
Superconducting Quantum Interference Device (SQUID)   102
    See magnetometer
Support for Science and Technology at Primary School level, *See* Accompagnement en Science et technologie à l'école Primaire (ASTEP)
surgery   56
surprise   4, 11, 30, 34, 56, 58, 96–97, 103, 111, 116, 126, 170, 193, 198, 203, 206, 208, 237, 241, 253
survey   x, 3–4, 12, 26, 136–137, 234–235, 240
surveyor's target   170, 179
sustainability   65, 75

INDEX

301

SV Tenacious (ship) 167
Swarte, Rieks 247
Swarts, Frédéric 152–153
Swarts, Théodore 152
Switzerland vii, 259, 263
syringe 73

tablet 257
Taborska, Małgorzata 123
Taché, Jean-Paul 84, 98–99, 104
tachistoscope 55
tacit knowledge 29, 31, 39
Talas, Sofia 123, 244
Tammeorg, Johannes 57
tangent 218, 220
Tanner, Adrian 180
Tartu (Estonia) x, 10, 50–52, 54–57, 59, 62, 64, 140, 158, 226
teacher 6, 10, 21, 50, 62, 85, 91, 95, 114–115, 120, 127–130, 141, 153, 159, 169, 171, 182, 185, 190, 212, 214–215, 237–239, 248
  education 21, 120
  training 6, 10, 120, 153, 169, 212, 237
  teaching aid 50, 52, 54, 57, 61, 63, 94, 240
teaching and learning 15, 33, 62, 158–159, 161, 163, 165, 167, 169, 171, 173, 175, 177, 179–180
teaching method 261
teaching of experiment 183–185
teadus 50
team 6, 42, 45, 73, 75, 87, 90–91, 95–97, 108, 110, 116, 118, 120, 161, 171–172, 179, 201, 203, 206, 208, 214, 257, 267, 269, 278
teapot 28
Technasia 261
Technical University Twente, Enschede, Netherlands 260
technique 5, 31, 61, 75, 79, 83, 85, 122, 130, 164, 170, 261, 279
Technolab, Leiden, Netherlands 257
technology vii, 2–3, 14, 27, 37, 44, 47, 61, 88, 91, 95–97, 105, 112–113, 115–116, 120–121, 124–126, 160, 162, 189, 203, 245, 250, 253, 259
telecommunications 203
telephone 57, 73, 105
  Bell and Siemens 105

telescope x, 16, 24–26, 31, 45, 85, 140, 158, 167, 169–171, 186
telos 17
temperature 86, 164, 192, 194
template 171–172
text-driven narrative 29
textual criticism 33
Teyler van der Hulst, Pieter 244
Teylers Foundation 244
Teylers Hofje, Haarlem, Netherlands 244
Teylers Museum, Haarlem, Netherlands 1, 4–5, 9, 13, 48, 81, 243–247, 249–254, 256, 258, 260, 262
  See Garden Observatorium, Haarlem, Netherlands
  See Lorentz Lab, Haarlem, Netherlands
  See Oval Room, Haarlem, Netherlands
theatre 7, 91, 181–182, 190, 198, 200–204, 206, 243, 247–248, 253, 260, 262, 279
theodolite 169
theology 67–68
theoretical 15, 25, 27, 58–59, 61–63, 140–141, 157, 184, 189, 213, 218, 258
  concept 27, 213
  physics 140
theory of general relativity 278
theory of relativity 247
thermocouple 151
thermodynamics 186
thermology 7
thermometer 42, 188
thermometry 42
thesis 8–9, 88, 105, 140–141, 149–152, 154, 157, 175–176, 188, 191, 240
Thollon spectroscope 87
Thomson, Joseph John 143, 188, 196–197
Thomson's tube 143, 188, 196–197
Thomson, Linda 60, 62
thought experiment 220
Thouin, Jérémy 85–86
three-dimensional view 79
thunder church 68
thunder house 4, 47, 186, 228–230, 233
time interval 266–267, 270, 273
Time Machine, The (film) 203
timed ignition 76
Tobin, William 92–93
Toneelschuur, De 247

toothed wheel   124
Torricelli, Evangelista   187, 193–195
torsion balance   31
tour   2, 4, 6, 95, 154, 170, 244
    guided   2, 80, 90, 95, 127, 154, 264, 279
traction   218
tractional motion   218–220, 223
tractrix   218–219
travel time   273
trial and error   40, 100
tricorn   36
trigonometry   172
trust   vii, 77, 172
Tsitou, Fay   199
tuning fork   85, 103
Tuscan   266
Tyldum, Morten   203
tympanum   176
Tyndall, John   182, 188, 202

UBO, *See* Université de Bretagne Occidentale
    (UBO), Brest, France
Ughent, *See* University of Ghent (Ughent),
    Ghent, Belgium
Uilkens, Jacobus Albertus   68
UMAC – International committee for
    university museums and collections
    98–99, 104
understanding   vii, 1, 9–10, 12, 17, 21, 29,
    33, 39, 62–63, 68, 72, 100, 103, 143, 155,
    159–160, 167, 170, 176, 180, 207, 214, 220,
    224, 226, 241, 259, 262, 264
unexpected   7, 160, 169, 194, 208
uniform rectilinear motion   273
uniformly accelerated motion   266
Union des Professeurs de Physique et de
    Chimie (UDPPC), France   92–93
Union of Professors of Physics and Chemistry
    (UDPPC) *See* Union des Professeurs de
    Physique et de Chimie (UDPPC)
United Kingdom   3, 27, 72–73, 75, 202, 207
United States of America   vii, 152, 267
UNIVERSEUM network   96–97, 104
Université de Bretagne Occidentale (UBO),
    Brest, France   10, 209–211
    See Cabinet of Curiosity, Brest, France
    See Institute for Research in the
        Teaching of Mathematics (IREM),
        Brest, France

See Scientific University College, Brest,
    France
Universities Centre for Cultural Students
    Activities (USVA), Groningen,
    Netherlands   83
University Claude Bernard Lyon   1
    (UCBL), Villeurbanne, France   8, 132–137
University College London, UL   58
University Hospital, Groningen,
    Netherlands   75
University Libraries Special Collections,
    Groningen, Netherlands   75
University Museum Groningen (UMG),
    Netherlands   4, 50, 56, 65, 67, 69–70,
    73, 77
University of Brest, Brest, France *See*
    Université de Bretagne Occidentale
    (UBO), Brest, France
University of Ghent (UGhent), Ghent,
    Belgium   8, 139–145, 147–149, 151–155
    See Museum for the History of Sciences
University of Groningen, Netherlands   65,
    82
    See Heymans Institute for Psychological
        Research, Groningen, Netherlands
    See University Libraries Special
        Collections, Groningen, Netherlands
    See University Museum Groningen
        (UMG), Netherlands
University of Massachusetts Boston Honors
    Program   180
University of Modena and Reggio Emilia,
    Reggio Emilia, Italy
    See Laboratorio delle Macchine
        Matematiche (Mathematical-
        Machines Laboratory) of Modena,
        Italy
University of Nantes, Nantes, France   88
University of Oldenburg, Germany   267
University of Rennes 1, France   5, 7, 84–87,
    91–93, 96–100, 102, 104
    See Beaulieu Scientific Campus
    See Culture and Heritage team
    See Culture-Heritage Service
    See Fablab
    See Faculty of Sciences of Rennes
    See Glass and Ceramics Laboratory of
        Rennes
    See Laboratory of Laser Physics

INDEX

See Laboratory of Physics of Infra-Red Spectroscopy
See Rennes Institute of Chemical Sciences
See Verres et Céramiques Laboratory
See École Supérieure d'Ingénieurs de Rennes (ESIR)
University of Tartu Art Museum, Tartu, Estonia   x, 10, 50–52, 54–57, 64
University of Tartu, Tartu, Estonia   x, 10, 50–52, 54–57, 62, 64
See Institute of History and Archaeology, Tartu, Estonia
See University of Tartu Art Museum, Tartu, Estonia
university scholar   vii
USB stick   96–97

vacuum   38, 66, 81, 153–154, 161–162, 186, 188, 192–193, 195, 197, 201, 203, 207, 264, 268, 271, 276–278
vacuum pump   38, 81, 154, 186, 188, 192, 207, 268, 271
boiling of water   192, 194
coin and the feather   192
extinguishing of flame   192
feather and glass ball   271
gradual disappearance of sound   192
Huygens   154
inflation of a sealed bladder   192
Magdeburg hemispheres   192
vacuum tube   161–162
valves   76
Van de Graaff generator   134
Van Heurck collection, Antwerp, Belgium   154
Varian mass spectrometer   102
Velox   152
Ven, Elisa Van der   247
Venus   167, 170, 177–179
Verne, Jules   181, 189
Verres et Céramiques Laboratory, Rennes, France   88
video   x, 2, 7–9, 12, 34–35, 37, 41–49, 76, 80, 86, 91, 99–100, 123, 135–136, 138, 161, 171–172, 190–191, 194, 199, 203
music   172
Violle actinometer   86, 91

Violle, Jules   86, 91
visceral   11, 19, 30–31
visual   7, 17, 28–29, 31, 45, 62, 190, 199, 253, 259
visual culture   31
vmbo -lower vocational, *See* school
voice   9, 14, 46, 78, 174
Voice Key (Schallschlüssel)   78
Volta pistol   199
Volta, Alessandro   187, 257
Voltaic pile   186, 255, 257
voltmeter   126, 151
volunteer   3–5, 30, 273–274
vwo – pre-university, *See* school

Wagentje van Stratingh (electric car)   65
Washington DC, USA   168, 170
watch (time-keeping)   4, 218, 266, 272–273, 278
water prism   240–241
Watt steam engine   94
Watt, James   94
weather   160, 194, 252
weight perception   77
Weiss, Pierre-Ernest   85
Wells, Herbert George   189
Western Regional Centre for Physical Measurements   102
Whale, James   203
whole-body   11, 22–23
whole-instrument   22–23
Wilcoxon test   236
Wilson cloud chamber   47
Wimshurst electrostatic generator   44, 68, 186
Wimshurst machine   111–112, 195–196
Windler, H.   56
Winterthur   28, 59, 61–62, 109
model   61–62, 109
Portfolio   28, 59, 109
protocol   28, 59, 61–62, 109
wissenschaft, *See* science
witness   7, 31, 189, 193–194, 199, 202, 207, 248, 263, 279
Wittje, Roland   8, 44, 47, 161
WND Conference 2019, Noordwijkerhout, Netherlands   260
working models   36

workshop   4, 53, 58–59, 63, 65–66, 78, 119, 134, 136, 141–143, 153–154, 161, 210–211, 238–239, 248, 257, 260, 263–264, 273–274, 276–279
Wow factor   260
Wright of Derby, Joseph   230

X-ray   9, 38, 81, 88, 101, 151, 207
    apparatus   101
    crystallography   88
    tube   38, 88
Xia, Summer   180

Yad Vashem Archives, Jerusalem   205, 208
Yang, Yan   180

Ye, Jinwen   180
YouTube   7–8, 34, 47–48, 135–136

Zénobe Gramme machine   132–134
Zimmermann, E. (Leipzig – Berlin)   77–78, 108
Zone d'Education Prioritaire (ZEP, Priority Education Sector)   130
zoology   95, 102, 139, 157

Ørsted, Hans Christian   89, 126, 196, 255, 257

Printed in the United States
by Baker & Taylor Publisher Services